Tourism, Growth and Sustainability

Rosaria Rita Canale · Rita De Siano

Tourism, Growth and Sustainability

Investigating New Strategies to Promote Growth

Rosaria Rita Canale
Department of Business
and Economics
Parthenope University of Naples
Naples, Italy

Rita De Siano
Department of Business
and Economics
Parthenope University of Naples
Naples, Italy

ISBN 978-3-031-85484-2   ISBN 978-3-031-85485-9 (eBook)
https://doi.org/10.1007/978-3-031-85485-9

© The Editor(s) (if applicable) and The Author(s), under exclusive license to Springer Nature Switzerland AG 2025

This work is subject to copyright. All rights are solely and exclusively licensed by the Publisher, whether the whole or part of the material is concerned, specifically the rights of translation, reprinting, reuse of illustrations, recitation, broadcasting, reproduction on microfilms or in any other physical way, and transmission or information storage and retrieval, electronic adaptation, computer software, or by similar or dissimilar methodology now known or hereafter developed.
The use of general descriptive names, registered names, trademarks, service marks, etc. in this publication does not imply, even in the absence of a specific statement, that such names are exempt from the relevant protective laws and regulations and therefore free for general use.
The publisher, the authors and the editors are safe to assume that the advice and information in this book are believed to be true and accurate at the date of publication. Neither the publisher nor the authors or the editors give a warranty, expressed or implied, with respect to the material contained herein or for any errors or omissions that may have been made. The publisher remains neutral with regard to jurisdictional claims in published maps and institutional affiliations.

Cover illustration: © Melisa Hasan

This Palgrave Macmillan imprint is published by the registered company Springer Nature Switzerland AG
The registered company address is: Gewerbestrasse 11, 6330 Cham, Switzerland

If disposing of this product, please recycle the paper.

Preface

This book originates from the intention to present, in a systematic way and with a critical perspective, the issue of tourism sustainability in search of new strategies to promote growth. The growth of the tourism sector, mainly due to the progressive market openness since the 1990s and the reduction of transport costs, has certainly brought great short-term benefits to both advanced and developing countries. However, as with any other productive sector, this expansion threatens to cause controversial effects on the environment and society in the long run.

Tourism, unlike other productive sectors, needs quality destinations to be consolidated. The sustainability of tourism is therefore not only linked to the need to preserve life on earth, but above all to the need to enable the preservation and development of the sector itself.

Taking this feature into account, tourism, if governed according to the principles of sustainability, can become a sector on which to focus not only to sustain growth while respecting the environment and the communities, but also to strengthen the resilience of destinations in the event of adversities.

Naples, Italy   Rosaria Rita Canale
                                                                            Rita De Siano

Acknowledgements

Thanks to Supraja Yegnaraman at Palgrave for her support and thanks the anonymous reviewers for their very thoughtful comments. Their observations were extremely useful and helped us to better focus on motivations of the book and to position it with respect to the existing literature.

Funding

Financed by the European Union—Next Generation EU, Mission 4 "Education and Research"—Component C2 CUP I53D23002850006. Project Title: "The role of skill-biased technological change in immiserizing growth, spatial polarization and product quality upgrading" 2022ZKA4PZ.

Contents

1	**Introduction**	1
	References	5
2	**Tourism Sector Dynamics and the Issue of Sustainability**	7
	2.1 The Evolution of Tourism Phenomenon	7
	2.2 The Role of Tourism for Economic Growth: A Macroeconomic Perspective	13
	2.3 The Issue of Sustainable Development: The Three Pillars	16
	2.4 Evolution of the Relationship Between Economic Growth and the Environment	18
	2.5 Interdependence Between the Economic System and the Environment (EKC): The Role of Tourism	22
	2.6 Sustainable Tourism: Advanced vs. Developing Countries	26
	References	28
3	**Tourism and the Environment**	33
	3.1 Tourism and Environmental Sustainability: A General Framework	34
	3.2 Climate Change and the Expansion of the Tourism Sector: The Role of Transports	38
	3.3 Tourism and Climate Action: The Conflict Between Poverty Reduction and Environmental Protection	41

3.4 The Innovation in the Tourism Sector as Strategy
     Against Climate Change   44
3.5 The Evolution of the Institutional Design About
     Tourism and Climate Action: Where Do We Stand?   48
References   51

4 **Tourism and Sustainability of Destinations: Some Analytical Tools**   55
4.1 The Complexity of the Phenomenon of Tourism
     Sustainability   55
4.2 The Evolution of the Tourism Product and the Life
     Cycle Approach from the Perspective of Sustainability   61
4.3 Sustainable Tourism and Market Equilibrium: The
     Effects on the Quality of Life   64
4.4 Tourism Sustainability and Collective Preferences   69
4.5 Concluding Remarks   74
References   76

5 **Contrasting Impacts of Tourism Expansion**   79
5.1 The Controversial Effect of Tourism on the Territory:
     The Phenomenon of Overtourism   79
5.2 Tourism Pressure Indicators: Alternative Perspectives   82
5.3 The Phenomenon of Overtourism and the Different
     Nature of the Territories   86
5.4 Resilience and Tourism   92
5.5 Recovery and Resilience: Definitions and Measurement   94
5.6 Reconciling Contrasting Impacts: The Critical
     Resilience Threshold of a Tourist Destination   101
References   103

6 **Tourism, Innovation and Sustainability in Europe**   109
6.1 The Dynamics of the Tourism Sector in the Main
     European Countries and the Issue of Sustainability   110
6.2 Tourism and Sustainability in the European Union:
     The Institutional Framework   113
6.3 The Certification System and the Regional Network
     in Europe   118

6.4	Innovation in Tourism as a Mean to Reach the Objective of Sustainability: The Case of Europe	122
6.5	The Future of Tourism in Advanced Economies: Limits, Opportunities and Challenges	126
Appendix		129
References		130

Index 133

List of Figures

Fig. 2.1	Temporal evolution of the regenerable resource stock (logistic function) (*Source* Own elaboration)	21
Fig. 2.2	Environmental Kuznets Curve (EKC) (*Source* Pettinger [2019])	23
Fig. 3.1	Bidirectional connection between tourism growth and environmental degradation (*Source* Own elaboration)	35
Fig. 3.2	Tourism direct and indirect environmental impact (*Source* Own elaboration)	37
Fig. 3.3	Tourism, environment and sustainable development goals (*Source* https://www.oneplanetnetwork.org)	38
Fig. 3.4	Carbon emission intensity per GDP and tourism innovation in Europe (2005–2019) (*Source* Own elaboration on Eurostat. Data presented in Canale and De Siano [2025])	47
Fig. 4.1	The complexity of the objective of sustainable tourism (*Source* Own elaboration)	59
Fig. 4.2	Tourism Area Life Cycle Model (*Source* Butler [1980, p. 7])	61
Fig. 4.3	Social equilibrium and market equilibrium in the tourism sector	66
Fig. 4.4	Common resources and the output of tourism firms	67
Fig. 4.5	Exploitation of tourist resources and quality of life	68
Fig. 4.6	Maximization of the net benefits of tourism and preferences of firms and residents	71
Fig. 4.7	Net benefits and alternative indifference curves	75

Fig. 5.1	Tourism density in selected European Countries (*Source* Own elaboration on Eurostat data)	87
Fig. 5.2	Tourism intensity in selected European Countries (*Source* Own elaboration on Eurostat data)	88
Fig. 5.3	Tourism territorial pressure (ml) in selected European Countries (*Source* Own elaboration on Eurostat data)	89
Fig. 5.4	Short-stay accommodation offered via collaborative economy platforms in selected touristic European cities (*Source* Own elaboration on Eurostat data)	90
Fig. 5.5	City tourism congestion index in selected touristic European capitals (*Source* Own elaboration on Eurostat data)	91
Fig. 5.6	Scale, change and resilience (SCR) in tourism (*Source* Lew [2014])	93
Fig. 5.7	Impact of a recessionary shock: region returns to pre-shock trend (*Source* Martin and Gardiner [2019])	96
Fig. 5.8	Impact of a recessionary shock: lack of resilience (*Source* Martin and Gardiner [2019])	97
Fig. 5.9	Impact of a recessionary shock: possible positive outcomes (*Source* Martin and Gardiner [2019])	97
Fig. 5.10	Resilience index for total stays after the pandemic by Covid-19 (*Source* Own elaboration on Eurostat dataset)	100
Fig. 6.1	Arrivals at tourist accommodation establishments in a sample of European countries (2007–2023) (*Source* Own elaboration on Eurostat datasets)	112
Fig. 6.2	Arrivals in selected years (*Source* Own elaboration on Eurostat datasets)	113
Fig. 6.3	Number of establishments in selected years (*Source* Own elaboration on Eurostat datasets)	114
Fig. 6.4	Arrivals at tourism accommodations per 1000 inhabitants in European countries (*Source* Own elaboration on Eurostat datasets)	114

List of Tables

Table 3.1	Global average CO_2 emissions per passenger-kilometre (PKM) travelled, 2016 and 2030 (kg)	39
Table 3.2	Emissions from international tourist arrivals by mode of transport and geographical area, 2016 and 2030 (Mt of CO_2)	40
Table 5.1	Resilience index using total stays at short-term accommodation in European cities	100
Table 6.1	ETC/ULS proposal for TOUERM indicators	129

CHAPTER 1

Introduction

Abstract This chapter introduces the reader to the debate on sustainable tourism from an economic perspective. Tourism, in fact, is a phenomenon not only contributing to the growth of the territories in which it is established but also generating several criticalities in terms of overexploitation of resources, alteration of places, removal of the population from the most attractive areas, commodification of local traditions and culture, to mention a few. This consideration gives rise to the interest in searching for strategies and tools ensuring that tourism is sustainable.

Keywords Tourism · Controversial effect of tourism · Sustainability

The tourism-growth nexus, when observed from a pure national accounts' perspective, is considered unequivocally positive. The increase in aggregate demand is supposed to induce an increase of production both for the tourist and interconnected sectors. The inflow of currency deriving from international arrivals is said to generate additional resources to be invested in domestic physical capital with positive direct and indirect effects on employment. However, to reconcile the expansion of tourism with sustainable development, several and controversial aspects need to be considered in search of new strategies to promote growth. The research and identification of what should be defined as sustainable, for whom it should be sustainable and at what level it should be sustainable is a complex route. The World Tourism Organization (WTO) defined as sustainable the *"Tourism that takes full account of its current and future economic, social and environmental impacts, addressing the needs of visitors, the industry, the environment and host communities"* (UNWTO, 2025). Sustainable tourism activities are therefore those that act in harmony with the environment, the community and local cultures so that they can benefit and not be victims of tourism development.

Until the 1960s tourism in Western economies was considered an activity capable of generating "clean wealth" but following its strong expansion and transformation into a mass phenomenon, awareness of the risks associated with its uncontrolled increase has matured. In the less developed countries, tourism has only gained economic importance in relative recent times. The excess of flows arriving from advanced countries and the limited tools to manage them, also due to the increasing domestic tourism promoted by governments to close the seasonality gap and thus create jobs and stability in the employment sector, risk compromising their identity and balanced development.

A degraded environment, due to excessive pollution and tourist flows at the limits of the carrying capacity of a site or territory, contributes to increasing the fragility of the destinations spreading its negative effects on connected territories and at global level through the vertiginous increase of maritime and air transports. Therefore, it becomes of the utmost importance to find forms of organizing tourism activities in accordance with the objective of sustainability, to know the critical thresholds it should not overcome and to encourage its resilience or the ability to adapt and renew in response to changes. This book proposes a reconstruction both from a theoretical and applied point of view of the connection between tourism, growth and sustainability with the aim of providing analytical tools to investigate issues at the centre stage of the policymakers' debate especially in advanced economies. This relevance concerns not only the contribution to growth, but also the attitude to respond to the new challenges that economic transformations pose to society in terms of environmental, economic and social sustainability. However, analysing all the implications—for example on biodiversity—of the expansion and construction of sustainable tourism models goes beyond the scope of the book and the competences of the authors.

The subsequent chapters are an attempt to present, in a systematic way, the issue of tourism sustainability and its connection with growth from an economic perspective. The reader will be able to understand, by using straightforward and rigorous analytical tools, the interpretative categories to consider in order to measure and evaluate tourism's contribution to growth and its impact on environmental transformation and climate change both at local and global level. The following questions will be addressed: is tourism always beneficial for growth? Which is its impact on the environment and climate change? What role does tourism have in developing and emerging economies for the reduction of poverty? Can advanced economies rely on tourism to foster their stagnant rate of growth, increase internal convergence and reduce inequality? Is there a trade-off between tourism as an engine of growth and environmental protection and climate change? Can tourism-innovation strategies promote environmental sustainability? Might tourism play a role in determining the capacity of a given territory to resist or to recover in presence of extraordinary negative shocks? The contribution thus presents the connection between tourism growth and sustainability as a complex phenomenon, to analyse what tools are needed to make the link virtuous, while few unambiguous answers are available.

The second chapter aims at providing a general overview of the role of tourism in the economic system and to define the issue of sustainability, focusing on how the evolution of the relationship between growth, tourism and the environment has progressively changed. In addition to presenting a systemic vision, the chapter devotes particular attention to the way in which the effects of tourism on the economic system and the environment can be expressed in advanced economies and in developing countries.

The focus of the third chapter is the connection between the expansion of tourism and environmental sustainability with a particular focus on the effects of international tourism expansion on climate change. In this regard, distances and transport play a special role. The conflict between the reduction of poverty and the increase of employment on one side and effects on the environment of tourism expansion on the other side will be treated in the light of the possible innovations to be implemented to transform the sector towards climate change reduction and the protection of the planet. A special focus on the changes of institutional design about the action suggested to promote an environmentally sustainable tourism will end this chapter. The intent is to piece together the steps forward (or backward) inside the international context accounting for the political and economic global "disorder" occurred in recent times.

The fourth chapter deals with the relationship between tourism and sustainability from a theoretical point of view. The nature of "public goods" of both the environment and tourism resources lead to the use of microeconomic tools to identify the connection between economic choices and the impact on society. A complex picture emerges in which tourism is sustainable if both the aspects related to the environment and the society are considered. The consequence of this complexity is that the "optimal solutions" depend on the context to which they are applied and cannot be qualified as universally valid.

Chapter 5 deals with contrasting impacts of tourism. Its growth and its peculiar features are source of multiple phenomena having both negative and positive impacts on the territory and the economy. It can be a cause of excessive congestion and at same time instrument of rapid recovery after a shock, because of its resilient nature. This chapter provides analytical tools to understand and measure the phenomenon of overtourism together with a theoretical description of the broad concept of economic resilience and its application to the tourism sector. The aim is to provide a framework to understand the trade-offs between the ability to contribute to recovery and growth on one side and the congestion of the territory on the other side due to the tourism sector expansion.

The last chapter proposes a practical perspective for examining the relationship between tourism and sustainability in advanced economies with a particular focus to Europe. In particular, the measurement system proposed by the European institutions is presented (it involves both the public and private sectors with the aim of building synthetic indicators capable of qualifying both structures and territories ETIS). The diffusion of this methodology, applied for some time by numerous national and international organizations, is proposed as a first step towards the establishment of rules and policy indications consolidated at European level to make sustainable tourism a real tool for territorial development. An in-depth examination of the role of innovation in the tourism sector in Europe will provide elements to evaluate its contribution to sustainable development and to economic growth in the countries under examination.

Despite the complexity of the sector and the multiple perspectives from which to analyse the relationship between tourism, growth and sustainability, it emerges that tourism has the possibility of playing a key role in defining a strategy for protecting the environment and the community. The reference framework to draw inspiration from is the one provided by the United Nations agency for tourism (UNWTO, 2024). It provides a defined analytical framework and effective tools to achieve the objective of sustainability; however, it does not represent a constraint for the economic policies of the various countries. A proliferation of strategies is transversally—across countries, territories and type of activities—implemented by private companies and public institutions, which form national and supranational networks built up on a voluntary basis.

It would be needed, to make tourism a tool for sustainable development, a qualitative leap towards an economic policy framework that does not leave the achievement of the sustainability objectives identified by Agenda 2030 solely to the awareness and goodwill of individual subjects.

References

UNWTO. (2024). *The Glasgow Declaration on climate action in tourism*. UNWTO. Retrieved February 5, 2025 from https://www.unwto.org/the-glasgow-declaration-on-climate-action-in-tourism

UNWTO. (2025). *Sustainable development*. UNWTO. Retrieved February 5, 2025 from https://www.unwto.org/sustainable-development

CHAPTER 2

Tourism Sector Dynamics and the Issue of Sustainability

Abstract This chapter aims at providing a general overview of the role of tourism in the economic system and to define the issue of sustainability, focusing on how the evolution of the relationship between growth, tourism and the environment has progressively changed. In addition to presenting a systemic vision, the chapter will devote particular attention the way in which the effects of tourism on the economic system and the environment can be expressed in advanced economies and in developing countries.

Keywords Origin of tourism · Sustainable development · Environment · Developed and developing countries

2.1 The Evolution of Tourism Phenomenon

The UN defines tourism as *"a social, cultural and economic phenomenon which entails the movement of people to countries or places outside their usual environment for personal or business/professional purposes"* (UN, 2010, p. 10).

Distinguishing tourism from any other mode of movement of individuals between different places is important to correctly measure the size of

© The Author(s), under exclusive license to Springer Nature Switzerland AG 2025
R. R. Canale and R. De Siano, *Tourism, Growth and Sustainability*,
https://doi.org/10.1007/978-3-031-85485-9_2

the phenomenon and its economic, environmental and social impact on the places visited and on the visitors themselves.

The aim of this book is precisely to describe the phenomenon of tourism by measuring its main economic implications that contribute to determining the well-being of resident populations, the perception of places by visitors, the exploitation of tourism resources and the opportunities for future development of destinations.

Reliable statistics on tourism are necessary not only to measure its contribution to the national economy and to design the most effective policies but also to design marketing strategies, strengthen inter-institutional relations and evaluate the efficiency and effectiveness of management decisions of companies involved in related activities.

In the International Recommendations for Tourism Statistics 2008 of the United Nations, tourism is defined as a subset of travel. *"A visitor is a traveller who goes to a major destination outside his or her usual environment, for less than one year, for any principal purpose (business, leisure or other personal purposes) other than to be employed by an entity resident in the country or place visited"* (UN, 2010, p. 10). Specifically, *"a visitor (domestic, inbound or outbound) is classified as a tourist if his or her trip includes an overnight stay, or as a day visitor (or excursionist) otherwise"*.

Travelling and moving from one place to another has always characterized human life. Early communities moved and explored distant territories out of necessity, in search of food or safe places to live. A few millennia later, ancient Mediterranean civilizations began to carry out more structured explorations. The Phoenicians, skilled sailors and traders, sailed across the Mediterranean and beyond establishing trade routes and founding new colonies. Although the main purpose of their voyages was to trade with other peoples, they initiated a process of mixing cultures, ethnicities and beliefs, thus fostering a willingness to accept diversity which is essential for the growth and development of any society. The travels and explorations of the Greeks, on the other hand, were rather driven by their intellectual curiosity. In fact, in addition to mapping the vast territories they reached, they were more interested in understanding the philosophy, science and arts of the places they explored.

The early explorations were not simple movements from one point to another, but real journeys of discovery, which paved the way for an increasingly intense interaction between different cultures and societies. However, they laid the foundations for the concept of travel as we know it today, where travellers follow "beaten paths", benefit from established systems of provision and generally are isolated from difficulties and dangerous contexts.

The history of tourism, understood as the activities of people "*spending time away from home in pursuit of recreation, relaxation, and pleasure, while making use of the commercial provision of services*", begins long before the coinage of the word *tourist* from the late eighteenth century. In the Western tradition, indeed, organized travel with supporting infrastructure and the presence of renowned destinations to visit can be found both in ancient Greece and Rome. The birth of "heritage tourism", aimed at visit historic sites of recognized cultural importance, dates to that period. The Roman elite, moreover, used to move from the city centre to pleasant places on the coast or on the surrounding hills to escape the uncomfortable life condition characterizing the city during the summer. These seasonal transfers allowed them to enjoy better climatic conditions but also to take care of their body and mind with stays in destinations equipped with resorts and spas. Two millennia have passed but these remain the main objectives of modern tourists, too. The roots of modern tourism are to be found in religious tourism, the oldest form of travel in history, whose protagonists were the numerous pilgrims on their way to sacred destinations, in search of their spiritual dimension and their desire to get closer to God.

One of the earliest centres of worship and pilgrimage in antiquity was the sanctuary of Apollo in the city of Delphi, where visitors flocked in great numbers bringing gifts and tributes. Jerusalem also has a great history of pilgrimages behind it. Considered the Holy City for Jews, Christians and Muslims, over the years it has been a destination for travellers from all over the world. Around the fourth century, pilgrimages to Rome also began, while a few centuries later, around the tenth century, another important destination began to establish itself: Santiago de Compostela. Pilgrimages were not only spiritual journeys, but also opportunities for cultural and commercial exchanges which brought the peoples of Europe closer. Later, at the Age of Exploration in the fifteenth century, courageous sailors like Christopher Columbus, Vasco da Gama and Ferdinand Magellan revealed the existence of new continents and helped broaden the horizons of the known worlds. The contact between Europe and the Americas, Africa and Asia, although sometimes dramatic above all for indigenous populations, paved the way for the first real process of globalization and had a lasting impact on political, economic and social dynamics worldwide.

The spread of maps and travel reports by travellers who faced impervious journeys by sea and land, further fuelled people's curiosity and imagination, creating a sense of wonder and a growing desire to explore. A modern concept of tourism was born, and travel became synonymous with discovery, adventure and cultural enrichment.

The era of the Grand Tour, which took place in Europe during the seventeenth and eighteenth centuries, represents another fundamental chapter in the history of tourism. Rather, it was a cultural phenomenon, initially reserved for the aristocracy and later extended to the emerging bourgeoisie, involving young people, especially men, from the wealthiest families who undertook long journeys across Europe to complete their cultural and social education. The Grand Tour was above all an educational and formative experience that could last months and sometimes years. The classic itinerary included stops in France, Germany, Switzerland, but above all in Italy, considered the cradle of art, history and classical culture. Cities such as Venice, Florence, Rome and Naples were obligatory destinations, places where travellers could immerse themselves in Renaissance and classical art, study Roman and Greek antiquities, and appreciate the works of masters such as Michelangelo and Leonardo da Vinci. The exposure to different cultures, languages and social customs contributed to the personal development of individuals strengthening their open-mindedness, tolerance and skills. The Grand Tour served as a bridge between different cultures, promoting an integration that influenced also the art and the literature of the time. It marked the beginning of tourism as a tool for education and cultural appreciation, in line with the modern concept of travel as an enriching and formative experience.

The Industrial Revolution in the eighteenth century, by introducing new modes of transportation, such as the railroad and steamboat, radically changed people's mobility. Railways had a profound impact in enhancing the ability to travel longer distances in short times, thus disclosing new frontiers for trade, business and, above all, recreational tourism. The emerging middle class, benefiting from more leisure time and improved economic conditions, explored many previously inaccessible destinations. Seaside resorts, mountains and natural attractions became popular with such a growth in flows that it gave rise to the phenomenon later termed mass tourism. At the same time, steamboats transformed maritime travel. Shorter length and increased comfort made transoceanic travel more attractive, both for migration and pleasure. This led to an increase in international travels, further expanding the horizon of tourism.

Industrial Revolution favoured also the development of tourism infrastructures such as hotels, train stations and ports making travel experience more affordable to an increasing number of individuals. Guided tours were developed, and tourist agencies sprang up to meet the growing demand for assistance from tourists. The first travel agency was founded in London in 1865 by Thomas Cook. The *"Thomas Cook & Son"* was able to go beyond English borders by offering European itineraries. To make travel simple and affordable, the agency took care of every aspect: thanks to agreements with hotels and restaurants, customers could stay at reduced prices and by providing special "circular notes" they made it even possible to travel abroad without changing currency.

Changes occurred during the Industrial Revolution contributed to transform tourism from an exclusive experience to an activity accessible for many. Technological progress reshaped the perception of time and space, making the world a more connected and accessible place and laying the foundation for modern global tourism. However, although characterized by the spread of mass tourism does not date back to this period characterized by the spread of jet airliner and cheaper energy, mass tourism had already made its appearance during the Industrial Revolution thanks to the introduction of railways and steam power.

The scale and participation in tourism have changed greatly over time, and the number of destinations has multiplied, as did the equipment, infrastructure and facilities, but motivation and behaviours have not changed. Domestic tourism, as predicted in the early 2000s by the World Tourism Organization, has also grown exponentially in countries such as China, India, Thailand, Brazil and Mexico. These countries are proof that domestic tourism is a growing sector also in several parts of the developing world. In these countries, travel for leisure, pilgrimage or business is no longer the exclusive prerogative of the upper classes but also includes the lower and middle classes. The rise of a middle class with reasonable wealth and disposable income and a strong desire to travel is certainly behind the growth of domestic tourism in these countries, but an important role is also played by government policies to seek an alternative and less exploitative form of tourism development than that dominated by multinational capital interests, pushing for greater community ownership of tourism enterprises. The objectives of domestic tourism development in these countries are slightly different from those of more advanced countries, namely poverty reduction and wealth creation promoting national unity and integration; enabling local people to benefit from government investment in tourism infrastructure, including national parks and reserves; expanding domestic tourism-induced investment; transfer resources from richer to poorer areas and communities; and to compensate for seasonal variations in foreign tourism, thereby supporting tourism demand and employment in the tourism sector (Mazimhaka, 2007). However, where governments prefer to promote international tourism rather than domestic tourism, this is because of the foreign exchange earnings that the sector generates for the host country.

The twentieth century saw an unprecedented expansion of international tourism flows. The introduction of commercial flight made long-distance travel possible in much shorter timeframes. Aviation opened new destinations around the world, making previously remote and inaccessible places reachable in just a few hours. Technological progress played an equally fundamental role also in travel organization and planning. Online travel agencies, review sites and booking platforms made travel planning more accessible and customizable, allowing travellers to create tailor-made experiences and transforming tourism in one of the most popular leisure-activity for most of the population.

Nowadays, tourism represents one of the world's largest industries, with direct or indirect linkages with many other sectors of the global economy. It has been recognized among the largest generators of income, wealth and development in many local and national economies (Garau-Vadell et al., 2018). Before the Covid-19 pandemic, tourism contributed for more the 10% of the global GDP and the previous decade was characterized by tourism sector growth rate far exceeding that of the world economy. International tourist arrivals rose sharply from 25.2 million in 1950 to 1.29 billion in 2023 and Europe was confirmed as the first destination with more than half (54%) of the total arrivals. Since 1990, growth in international arrivals has risen faster than the global average also in the Middle East and the Asia and Pacific regions (UNWTO, 2020).

Due to the role played in global income, world trade and employment, especially for weak components of the labour force (young people and women), tourism became a crucial issue in governmental macroeconomic policies around the world. Moreover, due to the recent fear about the detrimental effects in terms of overcrowding and environmental degradation caused by the extraordinary growth of the number of arrivals, the concerns of policymakers have even grown further (Albaladejo &

González-Martínez, 2019; Ehigiamusoe, 2020; Po & Huang, 2008). Tourism, indeed, can contribute to irreversible damage to the environment because caused by higher pressures on fragile ecosystems, more intensive use of resources like land, water and air, higher levels of pollution, erosion, deforestation and construction of infrastructure such as resorts and roads that may destroy natural sites and heritage.

In this view, sustainability and responsible tourism could be the only solution for an economic tourism-led growth path compatible with the preservation of destinations' resources availabilities.

2.2 The Role of Tourism for Economic Growth: A Macroeconomic Perspective

The tourism-growth nexus, if viewed from a short-term perspective focused solely on the monetary phenomenon, is considered unequivocally positive. The link between tourism and economic growth traces back to the seminal theoretical contribution by Butler (1980) on the evolution of tourist destinations (Tourism Area Life Cycle approach-TALC). Since then, following different methodologies and extending analyses to various time periods, scholars have shown that tourism development actually does contribute to the economic growth of a destination area (Balaguer & Cantavella-Jordá, 2002; Brida et al., 2016a, 2016b; Dogru & Bulut, 2018; Gunduz & Hatemi, 2005; Kim et al., 2006; Mérida & Golpe, 2016; Tugcu, 2014). There is a broad consensus on the positive contribution of tourism in boosting economic growth through channels such as production, employment and tax revenues. Tourism expenditure in each period, in all its components, acts on the local and national economy, determining a shock in demand for local products and services. In detail, this demand activates the sectors in which the expenditure is made in a direct manner and propagates in the local (and national) economy in an indirect manner through the links between sectors and, finally, in an induced manner through household income expenditure. Impacts may be distinguished as follows:

- direct impact: the impact generated on the demand for goods and services by the productive sectors involved in the activities engaged in the tourism sector.

- indirect impact: the impact determined by the increase in demand and supply in the supply chains activated.
- induced impact: the effect of the re-injection of labour and capital income into the economic system and the re-investment of tax revenues in the form of public expenditure.

The sectors that benefit most from tourism spending, both in terms of impact on GDP and in terms of employment, are the service sectors but considerable effects are also found in industry and in the primary sector (agriculture, fishing, forestry, mining, deposits).

However, the benefits of tourism can cover a wide range of aspects. They concern investments in new infrastructures (Sakai, 2009), human capital (Blake et al., 2006) and technology (Feng & Morrison, 2007; Lemmetyinen & Go, 2009) and efficiency improvements of local firms driven by higher competition and exploitation of economies of scale. Being considered an alternative form of export, tourism may represent prime source of foreign exchange earnings that by improving country's balance of payments helps to reduce its deficit. Social capital accumulation follows cultural exchanges between people of different geographical origins or backgrounds and, finally when aimed at promoting local natural resources, tourism may also represent an important channel for the protection of the environment and wildlife.

The overall contribution of tourism strongly depends on the ability of policymakers, institutions and stakeholders to create a sustainable model to manage the sector's growth. In fact, an uncontrolled increase of arrivals can compromise the sustainability of the tourism sector itself and, therefore, weaken its capacity to contribute to the growth of economies in the future. The United Nations World Tourism Organization (UNWTO) referred to this phenomenon as overtourism, that is *"the impact of tourism on a destination, or parts thereof, that excessively influences perceived quality of life of citizens and/or quality of visitors' experiences in a negative way"* (UNWTO, 2018, p. 4). This phenomenon has been recently exacerbated in many destinations due to the concurrence of several factors like low-cost airlines, cruise tourism and increasing short-term holiday rentals, the latter driven by the proliferation of online booking platforms (Postma & Schmuecker, 2017; Veiga et al., 2018).

In this regard, two streams of thought, albeit starting from different theoretical frameworks, also recognize the potential risks of overexploitation of key resources for tourism and the need to pursue the goal of sustainable growth. The first, based on the well-known TALC model, describes an S-shaped relation between tourist arrivals and time, suggesting that the congestion of the territory, presumably caused by overcrowding, not only reduces current tourism flows but compromises its contribution to future economic growth. The evidence of a non-linear tourism-growth nexus has been confirmed by several studies. For countries where tourism represents the main component of national GDP, for example, a kind of Dutch disease may take place. Excessive dependence on tourism for growth entails pitfalls that need to be reflected upon in good time. Firstly, tourism encourages a process of economic restructuring focused on low value-added activities, characterized by low skills and low productivity, linked to precarious and seasonal employment. The excessive development of tourism, indeed, by causing currency appreciation shows detrimental effects also on the general competitiveness, squeezing other sectors contribution to GDP and fostering an unhealthy mix between public and private sectors' activities (Capó et al., 2007). Po and Huang (2008), among others, demonstrate an actual difference in the correlation between tourism and economic growth depending on the level of tourism specialization: countries with a low or a high tourism specialization register significant positive effects on growth, while countries with an intermediate level of specialization do not show any significant effect. The second stream of taught is well described by the Johnston and Tyrrell (2005) theoretical model which reveal the presence of an inverted U-shaped relationship between environmental degradation caused by tourism and growth. According to this model, although benefiting from tourism in terms of higher income, employment and tax revenue (Haralambopolous & Pizam, 1996), destinations appear to suffer from disadvantages caused by uncontrolled arrivals, such as congestion, environmental degradation and noise (Mason & Cheyne, 2000). Ehigiamusoe (2020), for example, finds such a non-linear relationship between tourism and environmental degradation for African countries.

Although the linkage between tourism and economic growth is primarily positive for all countries, here are wide differences across countries and within each country depending on tourism specialization and real GDP per capita growth. The heterogeneous effects of tourism may therefore be attributed to multiple factors such as the relative weight of the tourism industry in the overall economy, size and openness of each economy, production capacity constraints, relevance of local businesses in the tourism industry and potential damages from tourism activities.

The latter can take the form of losses of competitiveness and excessive resources exploitation due to congested environments.

Within the same country, the tourism-economic growth link is not uniform over time but may change according to the phase of the economic cycle and the occurrence of shocks hitting directly or indirectly the sector itself. In this respect, when countries are strongly specialized in tourism, the link between tourism activity and economic growth appears to be more pronounced during deep economic downturns (Shahzad et al., 2017). Moreover, since tourism-enhancing policies are particularly beneficial during recessions, tourism is proved to play a strategic role in stimulating economic recovery (De Siano & Canale, 2024).

2.3 The Issue of Sustainable Development: The Three Pillars

When talking about sustainability, or sustainable development, we refer to a development model that can meet the needs of present generation without compromising the possibility of future generations meeting their own needs. This concept is based on a holistic approach which covers different aspects of life on heart and considers all social, environmental and economic impacts that could arise from every decision or action taken today. Sustainable development, in fact, goes far beyond the mere protection of the environment, requiring also economic well-being and the presence of an inclusive society to allow the satisfaction of all needs of a community, material as well as immaterial. Reducing the consumption of natural resources and the environmental impact of the economic-social system, for example, could contribute to achieve the objectives of economic efficiency and social cohesion but would not guarantee that future generations would have the same opportunities.

A pessimistic view of the relationship between human activities and the environment hypothesizes that economic growth and the uncontrolled exploitation of natural resources may cause an irreversible deterioration in the quality of the environment and the destruction of all natural resources, both exhaustible and regenerable. In this case, the very survival of humanity would be at risk.

In contrast, proponents of a more optimistic view believe that appropriate regulation of activities and deployment of resources can create the conditions for achieving the goal of a sustainable growth, i.e. a state of the economic system in which increased production is compatible with the preservation of the quality of the environment and the stock of natural resources.

From a historical perspective, the concept of sustainability was formulated at the first United Nations Conference on the Environment in 1972, but it has only really taken shape since 1987, when the publication of the so-called Brundtland Report ("Our Common Future") clarified the goals of sustainable development.

Sustainable development is characterized by three main principles:

- Economic sustainability, i.e. the ability of an economic system to sustainably generate income and employment to support the population. With reference to a given territory, this means the ability to produce and maintain maximum added value by effectively combining resources to enhance the specificity of territorial products and services, including landscapes, natural and cultural systems and intangible heritage. Economic sustainability allows for a resilient and thriving economy, serving present and future generations. This principle requires responsible resource management, sustainable business practices and the promotion of circular economies.
- Social sustainability, focuses on promoting equitable and inclusive communities where well-being, in terms of security, health, education, democracy, participation, justice, is equally distributed among classes and genders. Social sustainability is essential to build societies in which every person can prosper, regardless of her/him background or condition. This aspect of sustainability contributes to the empowerment of communities and the maintenance of local traditions and rights together with the respect to their own territory.
- Environmental sustainability, i.e. the ability to maintain the quality and reproducibility of natural resources over time. This principle entails the responsible use of natural resources, reducing waste and pollution and conserving biodiversity. Compliance with this principle requires a commitment to combat climate change, protect ecosystems and restore degraded environments.

The origins of the "three pillar" paradigm can be found in the intense debate that has characterized the institutional course of the sustainable development concept in recent decades, starting with the Brundtland Report of 1987. Subsequent events, however, have contributed to the realization that sustainable development and the conjugation of the three fundamental dimensions (social, environmental and economic) do not represent a static condition but rather a vision that can change over time, also because of the changing strategic interactions between them. In this regard, therefore, all actors, public and private, must not act in isolation or limiting themselves to the sector, but must take into account all changes that may alter the balance between the three dimensions of sustainable development. Sharing objectives should allow, in the long term, the attainment of higher levels of quality of life even in geographical areas where populations still live in extreme poverty, thus achieving a reduction in the differences between the north and south of the world.

2.4 Evolution of the Relationship Between Economic Growth and the Environment

The growing interest in the concept of sustainable development is a consequence of the critical change in the paradigm linking economic growth and natural environment. For centuries, ever since the birth of capitalism and the Industrial Revolution, the natural environment has been regarded as separate from mankind, a pool of resources that men could use to survive or exploit to expand their activities. The needs of individuals had priority with consequences on natural resources endowments in the short as well as in the long term completely disregarded. No concern or awareness of the changes that the growth of resource exploitation would have had on the quality of life of all living beings in the future. Rather, there was a belief that knowledge and technology would overcome the critical issues related to the development model adopted. In the late 1500s, the English philosopher Francis Bacon, indeed, prefigured the industrial society by focusing on technology as a tool to "tame" nature. He argued that "science should ensure man's dominion over nature, guaranteeing him better living conditions".

The exploitation of environmental resources accelerated with the First Industrial Revolution at the end of the eighteenth century and between the nineteenth and twentieth centuries with the Second Industrial Revolution. The first great change started in England, later expanding throughout Western Europe, marking an epochal shift of the society from a rural economy, characterized essentially by family-based organized agricultural activities, to a form of industrial activity that allowed to produce far greater quantities of manufactured goods than had ever been produced within the bottegas of craftsmen.

The invention of the steam engine (James Watt in 1769) initiated a technological advancement that not only increased production levels through the exploitation of thermal and kinetic energy but also made it possible to locate production sites near the urban centres where labour was most concentrated. With the growth and concentration of manufacturing activities, pressures on the environment also increased due to pollution and overexploitation of natural resources. The period between the First and Second Industrial Revolution, in fact, coincided with the most important colonial expansion of European countries in search of lands offering raw materials become scarce in the old continent, especially fossil fuels and metals.

The major concerns of companies and governments were usually related to the exploitation and management of natural resources, especially depletable ones. The only exception was the conservation of wild lands, which were taken care of through the establishment of parks or nature reserves. In this regard, Gifford Pinchot, responsible for forestry policies in the United States from 1898 to 1910 during the presidencies of William McKinley, Theodore Roosevelt and William Howard Taft, was firmly convinced of the need to preserve natural resources to guarantee their prolonged use over time. Despite the richness of timber in the country, Pinchot was among the first policymakers which aimed to achieve the goal of a socially efficient use of forests, at a time when they were still considered "inexhaustible". He contributed to the birth of the movement for the conservation of forests and natural resources in the United States.

Besides the overexploitation of natural resources, the increase of industrial production contributed to an acceleration of environmental degradation also through higher waste and polluting emissions. At that age, however, pollutant sources were also driven by intensive agricultural activities altering the soil with insecticides, pesticides and herbicides. All these sources of pollutants interfere with the natural functioning of ecosystems, altering their biogeochemical cycles and directly or indirectly causing loss of plant and animal diversities. Biodiversity, in addition to its intrinsic value, contributes also to determining the vulnerability of an area towards natural disasters and health diseases (almost half of the world's medicines are obtained from substances derived directly or indirectly from animal and plant kingdoms).

Biodiversity can be threatened by many factors like destruction, degradation and fragmentation of habitats attributable to natural disasters (fires, volcanic eruptions, tsunamis, floods, etc.) or other changes attributable to human activities. Biodiversity is increasingly threatened by overexploitation of natural resources, i.e. by levels of exploitation exceeding the regenerative capacity of environmental resources. Here, the problem is to identify the level of socially efficient (in the Paretian sense) exploitation of environmental resources, i.e. that trade-off between exploitation and preservation of resources that maximizes the net social benefits, given by the difference between social benefits and costs (damages), associated with the two activities. In fact, while the use of natural resources is essential for the expansion of the economic system and the increase in the well-being of individuals linked to economic factors, it can also be detrimental to the preservation of the environment, with negative effects on its ability to regenerate and, therefore, on the quality of life of individuals.

The economic system and the environment are strictly interdependent: the expansion of the first can cause irreversible alterations of the natural environment, and the latter can place serious limits on the expansion of the economic system due to its consequences on the quality of life. Policymakers, in this context, have the burden to identify strategies, as shared as possible, ensuring a sustainable use of environmental resources, guaranteeing future generations the same opportunities as current generations.

Natural resources can be useful and scarce, at the same time, hence the need for their efficient use according to the "Paretian principle", stating that there is no different allocation of resources that would increase the benefit of one individual without reducing the benefit of someone else.

The socially efficient choice depends on the type of resources. Resources, in fact, can be classified into exhaustible, as in the case of fossil fuels and minerals, and regenerable, as in the case of parts of ecosystems (animal populations, forests, ...) or environmental quality (water, air, soil). Regarding the former, as its current use reduces the stock available in future periods, an intertemporal optimization problem needs to be solved: the net marginal benefit of exploitation in the current period (difference between the benefits and co-benefits of present exploitation) must be equal to the present value of future net marginal benefits. If the benefits obtainable from future use are less important for present generation, the discount rate used to calculate the present value of future net benefits will be positive, implying a higher exploitation of the resource in the current period. For regenerable resources, on the other hand, sustainable exploitation requires that the use of resources in each period does not exceed their natural flow of expansion.

Regenerable resources can have a growth over time such as that shown in the graph in Fig. 2.1, where the x-axis shows the time and the y-axis the stock level of the available resource at any point in time. The curve shows the trend of a logistic function: for low stocks, the resource grows at increasing rates while, beyond a certain level, limiting factors in the environment begin to appear (in the case of animal populations, these could be for example the scarcity of food) and the resource continues to grow but at smaller and smaller rates until its biological equilibrium is achieved, i.e. the maximum quantity of the resource that can be reached in the absence of human withdrawal. The stock of a regenerable resources remains unchanged when the exploitation is equal to its expansion flow, which represents the achievement of the sustainable exploitation.

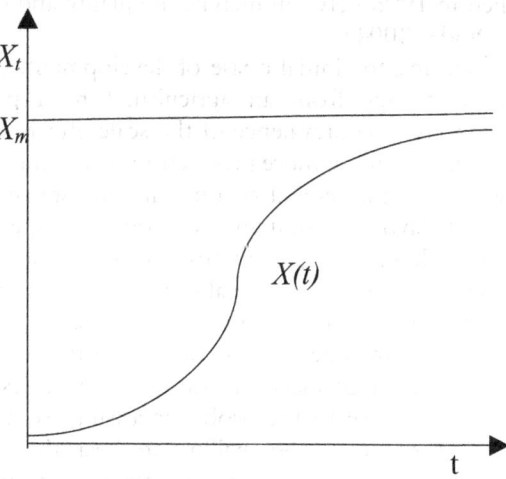

Fig. 2.1 Temporal evolution of the regenerable resource stock (logistic function) (*Source* Own elaboration)

2.5 Interdependence Between the Economic System and the Environment (EKC): The Role of Tourism

To the extent that countries have become aware of the environmental damage produced by their development models, the literature debate focused on the assessment of the environmental impact of economic activity and policies to support the recovery in terms of sustainability. The main evidence, driven by most of the empirical analyses, is that the quality of the environment deteriorates especially in the early stages of growth, because of a higher pressure on the environment (the level of degradation) essentially due to a scale effect, while, after a certain threshold, it recovers as income increases. This systematic relationship between income change and environmental quality can be represented by an inverted U-shaped curve, acknowledged as the Environmental Kuznets Curve (EKC) for the similarities with the relationship that Kuznets identified in 1995 between income inequality and economic development level (Dinda, 2004).

During the initial phase of development, brought about by the structural change from an agricultural to a predominantly industrialized economy, the prevalence of the scale effect makes the productive activity to pollute more: more production means more emissions. At a later stage, with the achievement of a mature and stable phase of economic growth, which favours a shift in the productive specialization towards activities with a lower marginal intensity of pollution, such as those related to the service sector, a substantial reduction in environmental damage may take place. Besides, at higher income levels, the decreasing marginal utility of per capita income and the increase in the social marginal cost of environmental damage make institutions, businesses and consumers themselves more sensitive to the problems arising from environmental degradation and, therefore, more willing to bear the costs of reducing pollutant emissions or exploiting natural resources respecting more sustainable use paths.

The more responsible approach to the natural environment stems from a greater social awareness of the damage that pollution can do to the health of individuals and the benefits associated with the preservation of natural resources. The level of social involvement that is achieved in this final phase, ranging from governments to companies, associations and finally individual citizens, favours the introduction of more restrictive environmental legislation and the adoption of more environmentally sustainable technologies and resources. The evolution of the socio-economic context favours the transition to a post-industrial phase, characterized by information and communication-based technologies that, thanks to lower levels of polluting emissions, help preserve the environment and improve human health.

Figure 2.2 shows graphically the relationship between the level of per capita income (x-axis) and environmental degradation/pollution (y-axis). In the initial phase of economic development and growth (phase 1), the trend of the Environmental Kuznets Curve is upward as the increase in production and the spread of industries have a strong negative impact on the environment, resulting in an increase in the level of degradation. However, what is of most interest to policymakers, businesses and consumers at this early stage is the general growth in employment, economic wealth and per capita income that follows the strong industrialization of the economy.

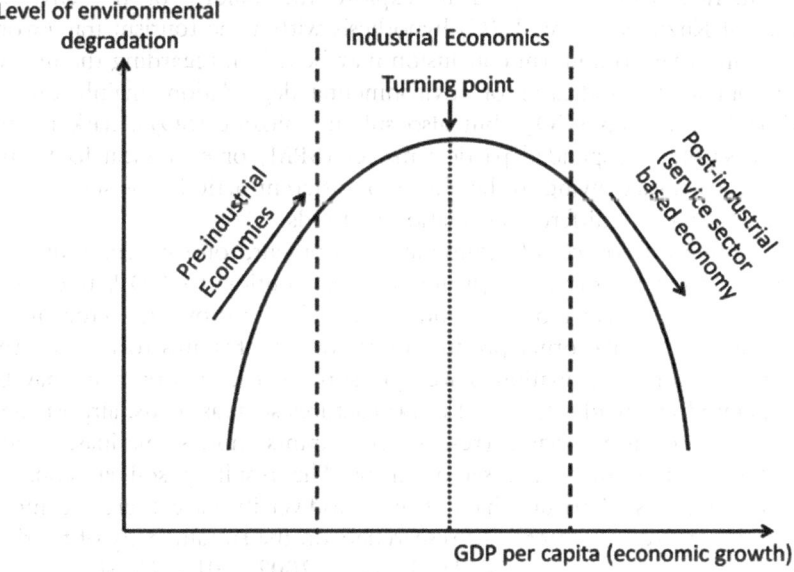

Fig. 2.2 Environmental Kuznets Curve (EKC) (*Source* Pettinger [2019])

The positive correlation between per capita income and environmental degradation, on the other hand, becomes negative when a high level of welfare is reached (corresponding to high per capita income), when environmental quality becomes a scarce good and society begins to be willing to spend part of its wealth on improving environmental quality. In fact, the preservation of the environment is also promoted by the increased efficiency of new technologies with reduced costs for pollution abatement and resource recycling. Thus, on the one hand, environmental policies capable of reducing polluting emissions from consumption and production activities are stimulated and, on the other hand, service sectors with high added value and low emissions grows. There is no shortage of examples of the relocation of highly polluting production activities, even to different countries or geographical areas, to reduce emission concentration levels in the place of origin. The turning point on the Kuznets Curve (point of maximum) is the one at which the relationship between environmental degradation and per capita income is inverted and the curve assumes a decreasing trend.

Many scholars attempted to explore the validity of the environmental Kuznets Curve (EKC) hypothesis within the tourism framework showing that no uniform conclusion may be driven regarding the impact of tourism the indicator of environmental degradation, mainly carbon dioxide emissions (CO_2) but also sulphur dioxide (SO_2), dark matter (fine smoke), suspended particle matter (SPM) or ecological footprint. The results vary owing to the choice of different periods, the selection of the country and different estimation methods.

Among the sources of economic development, tourism represents one of the most important regarding its contribution to GDP, trade and investment. It also represents one of the fastest growing sectors in all countries and, like other productive sectors, contributes to increase the environmental degradation. Large pressure on the environment may be generated by an adjustment of infrastructures such as roads, airports and different tourism services (resorts, restaurants, hotels, marinas, shops, …) needed to make the sector grow. The resulting soil erosion, air pollution, loss of natural habitat and biodiversity have therefore motivated an increasing interest of researchers on the sustainability of tourism activities (Dubois et al., 2011; Gössling, 2002, 2013; Gössling et al., 2005; Holden, 2009; Neto, 2003; Perch-Nielsen et al., 2010; Saenz-de-Miera & Rosselló, 2014; Sun et al., 2024; Tsai et al., 2014).

From this point of view, tourism can be anything but a low-energy and low-emission sector, as already demonstrated by Gössling et al. (2005) for France, the Rocky Mountain National Park in the USA, the Seychelles Islands in the Indian Ocean and the city of Amsterdam with reference to the early 2000s. This outcome has been confirmed at global level by Sun et al. (2024) who estimated tourism carbon footprints for inbound, outbound and domestic travel over a sample of 175 countries. The authors suggested to introduce demand volume thresholds to align global tourism effects with the Paris Agreement requirements. Moreover, the promotion of environmentally friendly technologies applied to tourism activities could also contribute to the conservation of natural resources and reduce the level of emissions.

Recent studies, on the other hand, have attempted to validate the EKC hypothesis by studying the "trivariate link" between tourism development, carbon emissions and income growth (Bella, 2018; Lee & Brahmasrene, 2013; Paramati et al., 2017). Bella (2018) analysed the long-run equilibrium relationship and direction of causality between tourism flows and pollutant emissions in France, considered among the world's top tourist destinations, controlling specifically for tourism-driven output growth. As a signatory to climate change agreements, France has introduced environmental conservation policies geared towards a decarbonized economy by the end of 2050, the sustainable use of resources in tourist destinations and succeeded in raising public awareness of environmental conservation. Particularly relevant among these initiatives are the promotion of eco-certification of hotels and private services dedicated to tourism (reducing water use and recycling waste); information campaigns to avoid energy waste; the implementation of new transport projects to strengthen electrified rail links between tourist destinations; "green passports" for tourists supporting biodiversity; the use of environmentally friendly modes of transport (car-pooling and car-sharing) and soft mobility in urban areas. The results in terms of carbon emissions of this study support the recent strategic plan proposed by French politicians.

Paramati et al. (2017) demonstrate the importance of the classification of countries by economic development level to understand the linkages between tourism, economic growth and carbon emissions. They show that the impact of tourism on carbon emissions is reducing much faster in developed economies than in developing economies, providing evidence of the EKC hypothesis on the link between tourism growth and CO_2 emissions.

Finally, being climate change a global problem, tourism as to be considered globally, too, with national and international families moving towards a sustainable use of resources in tourist destinations. Virtuous global practices can contribute to reduce the damaging effect of tourism on environmental degradation thus confirming the EKC even in the presence of tourism-led growth.

2.6 Sustainable Tourism: Advanced vs. Developing Countries

Due to its beneficial aspects, tourism is being strongly encouraged in most countries. The greatest benefits, however, seem to be brought to developing countries where tourism contributes to job creation and development of local infrastructures and services, to be used also for activities different from tourism. These countries, usually disadvantaged in all other economic sectors, are often blessed with the most important attracting resources such as warm climates, breathtaking landscapes and rich cultural heritages. Moreover, tourism-related professions, not requiring special skills, can be carried out also by low-skilled workers or workers belonging to more disadvantaged groups abundant in the domestic labour force. Beneficial effects could then spread in the whole economy thanks to positive spillovers towards related sectors such as agriculture, crafts, transport and services and aggregate demand sustained by revenues from the tourism activities. Therefore, for least developed nations, facing enormous obstacles in reaching even the lowest level of development, tourism could represent the fastest and most valuable source of foreign exchange to catch the path to development.

However, linking poor countries' development "exclusively" to the tourism sector may have different drawbacks due to the vulnerability of the sector itself. First, tourism demand is extremely unpredictable and any negative shock, affecting the economy, environment or socio-political context, may cause its collapse. Political and economic stability is therefore highly necessary not only to support demand, but also to guarantee the constant presence of investment by both local and foreign capitals. Second, additional economic activities should be provided to counteract not only declines in tourist flows, caused by unfavourable events, but also the seasonality that characterizes the sector's activities. Moreover, native populations could also suffer a depletion of natural resources (water, land, food and energy sources) and the exclusion from the distribution of wealth generated by tourism growth that, if not properly managed, benefits only wealthiest social classes without any improvement for weak destinations.

Pushing on an extensive development of tourism activities, however, may have potentially damaging effects on nature, societies and cultures everywhere, regardless of the country's level of development. Cases of overexploitation of natural resources, overcrowding and gentrification of tourist resorts are increasingly widespread so as to require drastic measures and changes in tourism development strategies to recover in terms of sustainability.

Sustainable tourism has been defined as *"any form of tourism development, management or activity that sustainably maintains the environmental, social and economic integrity and health of natural, historical and cultural resources"*. The UN World Tourism Organization further stated that sustainable tourism *"takes full account of its current and future economic, social and environmental impacts, addressing the needs of visitors, the industry, the environment and host communities"* (UNWTO, 2005).

These principles make it clear that the sustainability of tourism is not a prerogative of specific forms of tourism (rural, nature, solidarity, sports, ...) but affirm that any form of tourism must be sustainable for future generations. To achieve this goal, tourism development path should integrate tourism and environmental policies and benefit local communities, in terms of improvement in their quality of life, while accomplishing tourists' expectations, guaranteeing them high quality and satisfying experience.

If tourism development is aimed at these goals, the creation of lasting jobs, the cultural heritage preservation, the wildlife preservation and the landscape restoration will be ensured.

Sustainability as a long-term goal can only be relevant if it can gather the support of current beneficiaries. Rich countries, as capable of ensuring the satisfaction of the basic needs of the entire population, have the possibility of pursuing sustainable tourism policies that implies a socially efficient intertemporal exploitation of resources by postponing part of the benefits to future periods. Most developing countries, instead, show chronic and serious macroeconomic problems, such as high unemployment rates, rapid growth of the working-age population, high inflation and interest rates, a growing current account deficit rate and a growing debt-to-service ratio requiring urgent and effective actions (Tosun, 2001). The latter, therefore, unless they find additional sources of foreign exchange earnings and employment generation, will support any form of tourism development available to them, including unsustainable ones. The scarcity of domestic capitals imposes severe constraints on strategic decisions regarding tourism development. It may be difficult to introduce changes to pursue sustainability goals in tourism development. In these contexts, the role of the government and institutions is rather to build ad hoc tourism strategies to cope with the high bargaining power of international tourism firms and tour operators, on the one hand, and with changes caused by external factors on the other. Even in the most depressed areas, indeed, tourism infrastructures and services are often based on the standards of advanced countries to provide the mass of tourists with the comfortable environment they are accustomed. This is also the case in those areas where the local population has difficulty even meeting their basic needs, such as housing, education and health.

The empirical evidence, in several tourism destination areas, demonstrates that it is difficult to follow a sustainable tourism path or expect tourism to promote greater equality in the distribution of its benefits, especially when government and tourism policymakers do not have the strength to oppose the interests of their own community to those of foreign investors. At the same time, they are also conditioned by travel intermediaries dominating the flow patterns which are usually based, not in destination regions, but in the tourist generating countries.

References

Albaladejo, I. P., & González-Martínez, M. (2019). Congestion affecting the dynamic of tourism demand: Evidence from the most popular destinations in Spain. *Current Issues in Tourism, 22*(13), 1638–1652.

Balaguer, J., & Cantavella-Jordá, M. (2002). Tourism as a long-run economic growth factor: The Spanish case. *Applied Economics, 34*(7), 877–884.

Bella, G. (2018). Estimating the tourism induced environmental Kuznets curve in France. *Journal of Sustainable Tourism, 26*(12), 2043–2052.

Blake, A., Sinclair, M. T., & Campos Soria, J. A. (2006). Tourism productivity: Evidence from the United Kingdom. *Annals of Tourism Research, 33*(4), 1099–1120.

Brida, J. G., Cortes-Jimenez, I., & Pulina, M. (2016a). Has the tourism-led growth hypothesis been validated? A literature review. *Current Issues in Tourism, 19*(5), 394–430.

Brida, J. G., Lanzilotta, B., & Pizzolon, F. (2016b). Dynamic relationship between tourism and economic growth in MERCOSUR countries: A nonlinear approach based on asymmetric time series models. *Economics Bulletin*, *36*(2), 879–894.

Butler, R. (2015). The evolution of tourism and tourism research. *Tourism Recreation Research*, *40*(1), 16–27.

Butler, R. W. (1980). The concept of a tourist area cycle of evolution: Implications for management of resources. *Canadian Geography*, *24*(1), 5–12.

Capó, J., Riera Font, A., & Nadal, J. R. (2007). Dutch disease in tourism economies: Evidence from 20 the Balearics and the Canary Islands. *Journal of Sustainable Tourism*, *15*(6), 615–627.

Cater, E. A. (1987). Tourism in the least developed countries. *Annals of Tourism Research*, *14*(2), 202–226.

De Siano, R., & Canale, R. R. (2024). The role of tourism in European regions' economic recovery: A spatial perspective. *Tourism Economics*, *30*(8), 2021–2042. https://doi.org/10.1177/13548166241248679

Dinda, S. (2004). Environmental Kuznets curve hypothesis: A survey. *Ecological Economics*, *49*(4), 431–455.

Dogru, T., & Bulut, U. (2018). Is tourism an engine for economic recovery? Theory and empirical evidence. *Tourism Management*, *67*, 425–434.

Dubois, G., Peeters, P., Ceron, J. P., & Gössling, S. (2011). The future tourism mobility of the world population: Emission growth versus climate policy. *Transportation Research Part A*, *45*, 1031–1042.

Ehigiamusoe, K. U. (2020). Tourism, growth and environment: Analysis of nonlinear and moderating effects. *Journal of Sustainable Tourism*, *28*(8), 1174–1192.

Feng, R., & Morrison, A. M. (2007). Quality and value network. Marketing travel clubs. *Annals of Tourism Research*, *34*(3), 588–609.

Garau-Vadell, J. B., Gutierrez-Taño, D., & Diaz-Armas, R. (2018). Economic crisis and residents' perception of the impacts of tourism in mass tourism destinations. *Journal of Destination Marketing & Management*, *7*, 68–75.

Gössling, S. (2002). Global environmental consequences of tourism. *Global Environmental Change*, *12*(4), 283–302.

Gössling, S. (2013). National emissions from tourism: An over looked policy challenge? *Energy Policy*, *59*, 433–442.

Gössling, S., Peeters, P. M., Ceron, J. P., Dubois, G., Patterson, T., & Richardson, R. B. (2005). The eco-efficiency of tourism. *Ecological Economics*, *54*(4), 417–434.

Gunduz, L., & Hatemi, J. A. (2005). Is the tourism-led growth hypothesis valid for Turkey? *Applied Economics Letters*, *12*, 499–504.

Haralambopolous, N., & Pizam, A. (1996). Perceived impacts of tourism: The case of Samos. *Annals of Tourism Research*, *23*(3), 503–26.

Holden, A. (2009). The environment-tourism nexus influence of market ethics. *Annals of Tourism Research, 36*, 373–38.
Johnston, R. J., & Tyrrell, T. J. (2005). A dynamic model of sustainable tourism. *Journal of Travel Research, 44*(2), 124–134.
Kim, H., Chen, M., & Jang, S. (2006). Tourism expansion and economic development: The case of Taiwan. *Tourism Management, 27*(5), 925–933.
Lee, J. W., & Brahmasrene, T. (2013). Investigating the influence of tourism on economic growth and carbon emissions: Evidence from panel analysis of the European Union. *Tourism Management, 38*, 69–76.
Lemmetyinen, A., & Go, F. M. (2009). The key capabilities required for managing tourism business networks. *Tourism Management, 30*, 31–40.
Mason, P., & Cheyne, J. (2000). Residents' attitudes to proposed tourism development. *Annals of Tourism Research, 27*(2), 391–411.
Mazimhaka, J. (2007). Diversifying Rwanda's tourism industry: A role for domestic tourism. *Development Southern Africa, 24*(3), 491–504.
Mérida, A., & Golpe, A. A. (2016). Tourism-led growth revisited for Spain: Causality, business cycles and structural breaks. *International Journal of Tourism Research, 18*(1), 39–51.
Neto, F. (2003). A new approach to sustainable tourism development: Moving beyond environmental protection. *Natural Resources Forum, 27*(3), 212–222.
Paramati, S. R., Alam, M. S., & Chen, C. F. (2017). The effects of tourism on economic growth and CO_2 emissions a comparison between developed and developing economies. *Journal of Travel Research, 56*(6), 712–724.
Perch-Nielsen, S., Sesartic, A., & Stucki, M. (2010). The greenhouse gas intensity of the tourism sector: The case of Switzerland. *Environmental Science & Policy, 13*, 131–140.
Pettinger, T. (2019). *Environmental Kuznets curve.* https://www.economics help.org/blog/14337/environment/environmental-kuznets-curve
Postma, A., & Schmuecker, D. (2017). Understanding and overcoming negative impacts of tourism in city destinations: Conceptual model and strategic framework. *Journal of Tourism Futures, 3*(2), 144–156. https://doi.org/10.1108/JTF-04-2017-0022
Po, W. C., & Huang, B. N. (2008). Tourism development and economic growth—A nonlinear approach. *Physica A: Statistical Mechanics and Its Applications, 387*(22), 5535–5542.
Richardson, R. B. (2021). The role of tourism in sustainable development. In *Oxford research encyclopedia of environmental science.*
Saenz-de-Miera, O., & Rosselló, J. (2014). Modeling tourism impacts on air pollution: The case study of PM10 in Mallorca. *Tourism Management, 40*, 273–281.

Sakai, M. (2009). Public sector investment in tourism infrastructure. In L. Dwyer & P. Forsyth (Eds.), *International handbook on the economics of tourism*. Edward Elgar.

Shahzad, S. J. H., Shahbaz, M., Ferrer, R., & Kumar, R. R. (2017). Tourism-led growth hypothesis in the top ten tourist destinations: New evidence using the quantile-on-quantile approach. *Tourism Management, 60*, 223–232.

Sun, Y.-Y., Faturay, F., Lenzen, M., Gössling, S., & Higham, J. (2024, December 10). Drivers of global tourism carbon emissions. *Nature Communications, 15*(1), 10384. https://doi.org/10.1038/s41467-024-54582-7

Tosun, C. (2001). Challenges of sustainable tourism development in the developing world: The case of Turkey. *Tourism Management, 22*(3), 289–303.

Tsai, K. T., Lin, T. B., Hwang, R. L., & Huang, Y. J. (2014). Carbon dioxide emissions generated by energy consumption of hotels and homestay facilities in Taiwan. *Tourism Management, 42*, 13–21.

Tugcu, C. T. (2014). Tourism and economic growth nexus revisited: A panel causality analysis for the case of the Mediterranean region. *Tourism Management, 42*, 207–212.

UN Department of Economic and Social Affairs, Statistics Division. (2010). *International recommendations for tourism statisitics 2008*. Studies in Methods. Series M; No.: 83 Rev 1.

United Nations World Tourism Organization. (2020). *UNWTO world tourism barometer*. United Nations World Tourism Organization.

UNWTO. (2005). *Making tourism more sustainable: A guide for policy makers*, pp. 11–12. UNEP and UNWTO.

UNWTO. (2018). *Overtourism? Understanding and managing urban tourism growth beyond perceptions*. Madrid, Spain.

Veiga, C., Santos, M. C., Águas, P., & Santos, J. A. C. (2018). Sustainability as a key driver to address challenges. *Worldwide Hospitality and Tourism Themes, 10*(6), 662–673.

CHAPTER 3

Tourism and the Environment

Abstract The focus of the chapter is the connection between the expansion of tourism and environmental sustainability with a particular focus on the effects of international tourism expansion on climate change. In this regard, distances and transport play a special role. The conflict between the reduction of poverty and the increase of employment on one side and effects on the environment of tourism expansion on the other side will be treated in the light of the possible innovations to be implemented to transform the sector towards climate change reduction and the protection of the planet. A special focus on the changes of institutional design about the action suggested to promote an environmentally sustainable tourism will end this chapter. The intent is to piece together the steps forward (or backward) inside the international context accounting for the political and economic global "disorder" occurred in recent times.

Keywords Tourism · Environment · Direct and indirect impact · Transports · Poverty · Institutional settings

3.1 Tourism and Environmental Sustainability: A General Framework

The environment is the most precious resource we have. Ecosystems provide support for life. It should be a priority to establish criteria within which to allow the development of human activity (Gössling, 2002; Gössling & Hall, 2006). In the discussion among scholars, economic activities, their development and contribution to growth often occupy the foremost place in the hierarchy of values. Preserving ecosystems and the environment, therefore, does not always find an adequate position in the scale of objectives, due to the existence of a trade-off between economic growth and safeguarding employment on the one hand and conservation of environmental heritage on the other. This trade-off does not only occur between the present and the future, but often between different areas of the world engaged in mutually subtracting resources without having a global and far-sighted vision and the awareness of belonging to the same ecosystem, whose elements interact even thousands of miles away. The current state of the earth with regard to progressive environmental degradation indicates that the inverted U-shaped Environmental Kuznets Curve (EKC) described in the previous chapter is an often imprecise tool for analysing the relationship between per capita income growth and environmental degradation: imprecise because advanced countries, whose per capita income is higher, do not always choose the path of mitigating the effects of expansion on the climate and imprecise because in a highly interdependent world it is not possible to set geographical boundaries in identifying the effects of progressive industrialization on the climate.

Considering the possible connections between per capita income growth and environmental degradation, tourism represents a further element of disturbance, since on the one hand the sector needs a preserved environmental context in order to develop, but on the other hand it exerts effects that are not in line with sustainability objectives if expanded in an uncontrolled manner (Torres-Díaz et al., 2024).

Figure 3.1 shows the bidirectional relationship between growth in the tourism sector and environmental degradation. The simple representation indicates that the growth of tourism causes degradation of the territory (the arrow that moves from left to right) which in turn leads to a reduction in tourist flows due to the transformation of the destination into an area no longer suitable for attracting visitors (the arrow that indicates the

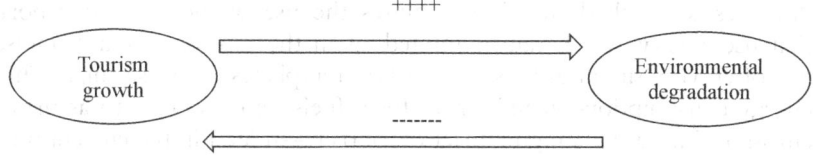

Fig. 3.1 Bidirectional connection between tourism growth and environmental degradation (*Source* Own elaboration)

movement from right to left). See the first part of Chapter 4 for further information on the phenomenon of overtourism.

As it is well known, tourism has grown worldwide over the last 20 years (excluding the Covid-19 setback). From 2010 to 2019 alone, arrivals worldwide went from around 973 million to 1465 million (an increase of over 33%). After the exogenous shock of the pandemic, in 2023 tourist flows in the world almost reached the pre-Covid-19 level (1305 million): if we account for the international turbulences and the interruption of relations between some countries in the world, this achievement is even more surprising. However, the expansion of tourism brings with it a series of consequences on the environment that cannot be underestimated and are under the scrutiny of international institutions (IPCC, 2022, 2023). These impacts can be classified as direct or indirect.

Scholars identify at least four direct areas on which the development of tourism activity impacts (Gössling, 2002; Gössling & Peeters, 2015; One Planet Sustainable Tourism Programme, 2021, 2023).

Tourism activities directly affect land use due to the need to use the territory for bathing establishments, mountain skiing facilities, for the construction of accommodation, for the services to be guaranteed such as parking lots, roads, airports, shopping areas. Accommodation facilities need surface land to expand and require additional services that erode further portions of the territory such as spaces dedicated to recreational activities. The destination of the territory for tourism activities alienates residents from the care of the places and generates potential factors of further degradation.

Tourism affects the use of energy and the volume of polluting emissions. Energy is used to support structures and activities intended for visitors who would use much less if they remained in their places of residence, but above all if they did not move. In fact, the need to cover distances to reach destinations requires the use of means of transport that use energy, which can be limited, as in the case of cars and trains, but often very significant, as in the case of airplanes or cruise ships. This energy consumption—mainly from fossil fuels—corresponds to as many emissions indicating a high impact of tourism—at least in the current way of managing it—on global warming.

Tourism leads to the use of a greater quantity of water and food consumption, perhaps in territories that do not have such resources in abundance, also generating distortions in the international distribution of resources. Water and food, resources so scarce in the peripheries of the world, reach destinations far from the needs of the resident population and undergo distortions in price and quantity that are unsustainable for the local market. Tourism affects life in rivers and seas and reduces the production of food that comes from them. What has just been described can be classified as a direct impact.

Even the indirect impacts on the environment should not be underestimated: the excessive development of tourist activities distorts the local organization of production by replacing—in the case of less developed countries—activities that have a lower environmental impact. The settlements on the territory of the local population are replaced by tourist facilities, causing damage to the territory and greater economic costs for the population. International tourism causes health risks through the exchange and spread of diseases that could have remained in a restricted area. Finally, tourism pressure on the environment is also revealed through waste production and congestion of sewage management facilities. Figure 3.2 schematically summarizes what has just been discussed, allowing us to have an overall picture of the impact of tourism on the environment.

Aware of the risks arising from the development of tourism activity, the United Nations Agenda 2030 has explicitly linked the tourism sector to environmental sustainability objectives (as well as to sustainability objectives as a whole) through the "Glasgow Declaration on Climate Action in Tourism" in collaboration with UNWTO within the framework of the Sustainable Tourism Programme of the One Planet network, of the UN Environment Programme (UNEP) (UNWTO, 2024; WTO, 2024).

Tourism that respects the environment must be inspired by responsible consumption and production models, reducing food and raw material waste. The model is that of the circular economy that minimizes waste from consumption and production and maximizes its use

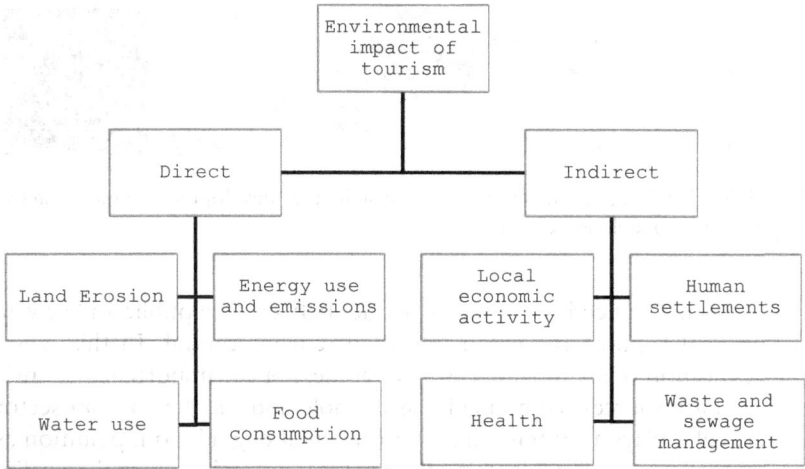

Fig. 3.2 Tourism direct and indirect environmental impact (*Source* Own elaboration)

as new raw materials for the next production cycle. In the tourism sector, being inspired by responsible consumption models also means reaching destinations that encourage values of environmental sustainability and are active tools for protecting nature, which respect life below water, life on land and biodiversity (target 12—Responsible consumption and production, target 13 climate action, target 14 life below water, target 15 life on land). These objectives can be achieved by measuring, according to the criteria suggested by the UNWTO, everything that can affect the objectives, decarbonizing tourism production (infrastructures, transport, accommodation facilities, etc.), supporting nature's ability to regenerate, but above all promoting collaboration paths between stakeholders interested in operating in the market. In this context, public governance and subsidized financing tools for companies that actively participate in achieving the objectives are central (target 17 partnership for the goals) (https://www.oneplanetnetwork.org/programmes/sustainable-tourism/glasgow-declaration).

Figure 3.3 reproduces the sustainable development goals to which environmentally friendly tourism can directly contribute.

It is clear that the objectives are closely linked to each other and that they can intersect with different weights depending on the nature of the

Fig. 3.3 Tourism, environment and sustainable development goals (*Source* https://www.oneplanetnetwork.org)

destination. The need to activate synergies between the public and private sectors that support the green transition remains central. In this sense, the acquisition of an ever-increasing awareness of the importance of environmental resources for humanity as a whole and for the tourism sector in particular plays a central role. Environmental degradation, pollution as well as land congestion, whatever their nature, not only reduce the quality of tourism today, but also compromise the existence of the sector in the years to come.

3.2 Climate Change and the Expansion of the Tourism Sector: The Role of Transports

Tourism is premised on travel. Tourists need to travel to reach the destinations. The travel will be longer the further away the destination to be reached. In turn, the different distances of destinations imply the use of different means of transport.

The means of transport that a traveller can use range from bicycles to cars, from trains to ferries, from ships to planes and so on. The further away the destination and the shorter the time employed, the more challenging, from an environmental point of view, the means of transport used.

With the expansion of tourism starting from the middle of the last century, not only the number of travellers has risen, but increasingly distant destinations have been reached. The change in lifestyles and the ever-increasing frenetic pace of daily life have reduced holiday time and made it necessary to reach destinations quickly. Globalization, which has seen a progressive acceleration since the 1990s, together with the exchange of goods and financial resources, has also pushed international travel and made very distant destinations increasingly attractive.

These phenomena, together with a general reduction in transport costs, have caused a strong expansion of tourism that contributed in turn to the expansion of the GDP of even the poorest and least accessible countries, but that also generated a growing use of means of transport covering long distances or cross large bodies of water, such as airplanes or cruise ships, responsible for high CO_2 emissions.

Tourism is therefore recognized as one of the productive sectors generating large amounts of polluting emissions and, therefore, it is under the scrutiny of international institutions. The UNWTO together with the International Transport Forum (2019) has thoroughly analysed the phenomenon with the aim of quantifying it and projecting the results into the future.

The analysis is conducted through a three-stage method that first estimates the demand for national and international tourism, then evaluates how tourists reach the destination and finally measures, also on the basis of the probable evolution and diffusion of technologies allowing the reduction of pollution, the potential CO_2 emissions (for details on the survey methodology see UNWTO & ITF, 2019). Unfortunately, the estimate excludes the contribution to polluting emissions from ship travel, due to the lack of data.

Table 3.1 shows the global average per passenger-kilometre (PKM) of CO_2 emissions, comparing the actual values of 2016 with those, obtained through the proposed estimation method, of 2030.

For each means of transport (car, bus, rail or air) and for all transports as a whole, PKM decrease, suggesting at first sight a reduction in emissions from 2016 to 2030.

However, if we combine this information with the evolution in tourist demand, the growing number of international trips and the covering of

Table 3.1 Global average CO_2 emissions per passenger-kilometre (PKM) travelled, 2016 and 2030 (kg)

Mode of transports	CO_2 emissions per PKM	
	2016	2030
Car	0.1136	0.0752
Bus	0.0300	0.0244
Rail	0.0205	0.0188
Air	0.1042	0.0798
All modes	0.0930	0.0680

Source UNWTO & ITF (2019, p. 36)

longer distances, the results change completely and instead point to a worrying picture (Table 3.2).

Emissions from air transport are supposed to grow everywhere in the world. Looking at the variations in the areas that include higher income countries, the 2016 levels are higher than the others (Europe, Americas and Asia and the Pacific) and the increase expected for 2030 is even higher (+48 Mt of CO_2 for the Americas and +57 Mt of CO_2 for Asia and the Pacific). The greatest growth in pollution from air transport is in Europe with +85 Mt of CO_2. Africa and the Middle East polluted little in 2016 and will produce a very limited increase in CO_2 (+16 Mt of CO_2 for Africa and +12 Mt of CO_2 for ASIA) in 2050 compared to other destinations. The other means of transport, rail, bus and car make a small contribution to CO_2 emissions in 2016. Especially for the areas that include advanced countries, this estimated contribution for 2030 is supposed to decrease. This reflects technological advances in transport that are pushing towards green transformation and greater use of alternative energy sources (electric cars for example). The distant values between the individual boxes (see especially the distance of Africa and the Middle

Table 3.2 Emissions from international tourist arrivals by mode of transport and geographical area, 2016 and 2030 (Mt of CO_2)

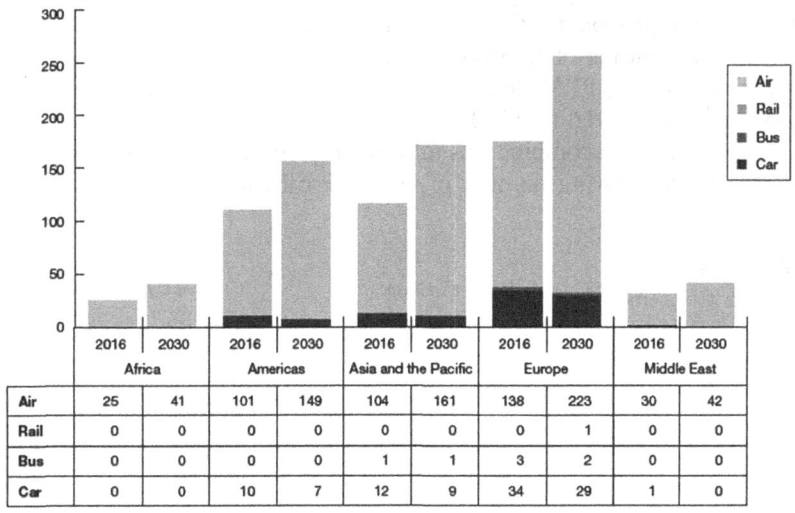

	2016 Africa	2030 Africa	2016 Americas	2030 Americas	2016 Asia and the Pacific	2030 Asia and the Pacific	2016 Europe	2030 Europe	2016 Middle East	2030 Middle East
Air	25	41	101	149	104	161	138	223	30	42
Rail	0	0	0	0	0	0	0	1	0	0
Bus	0	0	0	0	1	1	3	2	0	0
Car	0	0	10	7	12	9	34	29	1	0

Source UNWTO and ITF (2019, p. 40)

East from advanced countries) reflect the different habits of residents, the development of transport networks and above all the stage of evolution of the tourism sector in different areas of the world. The historical and almost permanent instability of the Middle East should not be neglected.

Although the estimation model does not provide certainty of the values presented, it clearly emerges that air travel is the most polluting means of transport, while train is the least polluting one, regardless of the geographical area considered. The expansion and growth of tourism—a phenomenon that is taking on great strength especially in Europe—contributes significantly to pollution and climate change, suggesting the adoption of alternative development models (Gössling & Peeters, 2015).

In fact, the evolution of the tourism sector can no longer depend only on the decarbonization strategies of the sectors involved, but, while waiting for new inventions to be realized, it must rather aim at the separation between the growth of the sector and the growth of emissions through tools and strategies that increase awareness of the phenomenon and involve all, national and international, stakeholders in the implementation of mitigation and adaptation strategies (UNWTO & ITF, 2019).

3.3 Tourism and Climate Action: The Conflict Between Poverty Reduction and Environmental Protection

The goal of limiting climate change and reducing poverty is recognized by international institutions as priorities. Both are part of the 2030 Agenda as fundamental pieces to ensure sustainability in a broad sense. The tourism sector, as well as other sectors, is involved in achieving these goals. The peculiarity of tourism, however, is that, as indicated in the previous paragraphs, the environment represents a necessary resource for its expansion, but at the same time it generates, especially through the growth of its international component, CO_2 emissions, which risk compromising its development. The expansion of tourism is also, if well managed, a powerful tool for combating poverty. Despite the structural changes, affecting all sectors of production due to the evolution of technology, tourism remains a highly labour-intensive sector. The percentage of employment in the tourism sector exceeds the contribution to GDP in both advanced and poor countries (OECD data available at https://data-explorer.oecd.org/). The tourism sector is resilient (see Chapter 5), as it is capable of recovering quickly after exogenous shocks, because of it is low capital-intensive nature.

It is therefore clear that, particularly in the tourism sector, the simultaneous achievement of the two objectives of combating poverty and environmental protection presents some challenges in their ability to be reconciled.

The institutional and policymakers hierarchy of values often puts economic interests first over those of environmental sustainability and prefers—following the consensus mechanism—current objectives over future ones. But this trade-off in recent times is becoming less and less evident following the growing awareness that these two aspects of collective life are inextricably linked.

According to the taxonomy proposed by Lankes et al. (2024) in identifying the evolution of the relationship between climate action and poverty reduction it is possible to identify two approaches to this issue (Sen, 1979):

1. The Utilitarian/Welfarist approach focuses its attention on the incremental advantages and disadvantages of individual utilities involved in the transformation process. This approach concludes that preferring one objective rather than another depends on the sum of the utilities of the individual, always measurable and quantifiable. While accepting the presence of catastrophic events that can compromise the logical structure of the analysis, ethical tools that approach the trade-off with a systemic vision are left aside.
2. The Rights and Justice approach, inspired by the contribution of Sen (1979, 2009), identifies the "right to development" as the main prerequisite for any policy action. The limitation of the "right to development" mainly concerns the poorest countries that suffer the consequences on climate change of the industrial expansion of rich countries, but also the weakest population groups in advanced countries that face price increases of necessities, like food and shelter, generated by negative shocks on climate. The Rights and Justice approach, although adopting a systemic perspective, requires as a premise the existence of values, which are difficult to define objectively.

These two approaches, providing both important tools, underline the existence of a trade-off to be solved according to the perspective considered more effective in managing it.

However, following Lankes et al. (2024) further evaluation tools to identify the connection between climate change and poverty growth do exist suggesting a relationship of the same sign, thus denying the existence of the trade-off. In fact, climate change limits the ability to produce income—think of the effects of environmental damage on the attractiveness of a destination—and damages for a long time precisely those populations that have fewer resources for refreshment and less adaptive capacity. Economic costs increase, agriculture reduces its productive capacity, and general health conditions worsen, generating amplified effects on the weakest segments of the population. Therefore, there would be no conflict between the objective of environmental protection and the objective of poverty reduction: accounting for possible climate changes in the construction of economic models reduces market failures, adding elements of realism to the analyses. The improvement in the quality of life achieved with policies to combat climate change contributes to the reduction of poverty more than the increase in GDP (Bruckner et al., 2022; Hubacek et al., 2017) through the creation of new jobs in previously little-explored sectors, which amplify the effects on local growth and promote social inclusion.

The elimination of this trade-off, however, depends on the policy strategies pursued, in the awareness that actions to combat climate change can no longer be postponed. The issue of the trade-off, for example, is related to the conflict between local development and global climate change: less developed countries that benefit from long-haul tourism may be more affected by policies that reduce flights between rich and poor countries with the objective to reduce air-transports emissions. However, only half of these less-developed countries would suffer significant damage from the decline in international tourism, which could be compensated by transfers and compensatory measures by international institutions or advanced countries (Peeters & Eijgelaar, 2014).

For less developed and poor countries it is particularly important to note that the fight against climate change, if managed in a balanced way, might generate a reduction in poverty and positive spillovers for advanced countries: the reduced industrialization of these areas of the world shows the existence of easier paths for the construction of new ways of carrying out tourism activities that are, from their establishment, compatible with environmental sustainability. The limited availability of capital, however, makes them dependent on foreign investments and the influx of resources from abroad: it is therefore necessary, together with cooperation at a global level, to have strict local regulation that makes forms of eco-friendly tourism attractive for the territory and a general international regulation that commits financial resources to not only seek short-term profit. An integrated approach between greater investments, sharing of strategies and close collaboration between advanced and developing countries, together with labour protection tools that reduce the size of the informal economy in this sector is needed (ILO, 2019).

3.4 The Innovation in the Tourism Sector as Strategy Against Climate Change

The relationship between tourism and the environment appears controversial: on the one hand, the expansion of the sector implies, especially through the growth of air transport at an international level, an environmental impact that needs to be remedied; on the other hand, environmental conservation is a premise for tourism. Innovation plays a central role in this controversial relationship.

Innovation is a tool for acquiring greater competitiveness through the creation of new development paths by companies or the sector implementing them and, depending on the content, it can represent a strategy to promote the fight against climate change.

The definition of innovation dates back to the seminal Schumpeter (1911 and 1934) contribution: it is depicted as "the introduction of a new product or bringing a different feature of an existing product to fore, the development of new techniques during the production process, and the availability of new resources in the supply chain". Innovation is concerned with the introduction of a new technology that allows the creation of a new product or the introduction of a new production process or the organizational structure of a company; it can be disruptive innovation, when it is introduced into an existing market or radical innovation when a new technology is created for the birth of a completely new product or it can be incremental innovation when it introduces new efficiency criteria without radically changing the structure of production (OECD, 2013; Smith, 2009). However, there is a form of innovation that goes beyond this technical classification and that arises in response to new needs of society not finding adequate response in the institutions.

In defining the role of innovation in tourism and its role in combating climate change, the UNWTO comes to our aid by including the path of renewal of the tourism sector within the sustainable development strategy: "Innovation in tourism, as elsewhere, is collaborative action between governments, academia, corporations, micro, small and medium enterprises (MSMEs) and start-ups, investors, supporting business partners (accelerators, incubators, etc.) and other stakeholders. Fostering a successful tourism innovation and entrepreneurial ecosystem requires connecting all stakeholders to collaborate opportunities and prioritizing capacity building in tourism and technology" (https://www.unwto.org/what-is-tourism-innovation).

A path of innovation in tourism that spreads its positive effects on the environment can be reconstructed as follows: research provides structures and models of sustainable innovation, contributing to the knowledge of the phenomenon. Companies apply discoveries to the organization and expansion of new activities: large companies develop technology and invest in order to achieve greater competitiveness, while smaller companies contribute to the diffusion and greater awareness of the contribution of innovation. Pursuing the goal of innovation requires the collaboration of governments and public institutions that promote and support the adoption of new technologies and new organizational choices.

However, the innovation process in the tourism sector is usually undersized because of its labour-intensive nature and the small firms belonging to the sector. In the present scenario calling for new strategies for combating climate change innovations often become a matter of small steps, rather than a matter of technological revolution. One innovation leads to another, producing slightly improved products and more efficient processes, and the innovation process becomes just another component of the investment process. Environment innovation in the tourism sector—i.e. innovation that results in a reduction of environmental impact, optimizing the use of resources might be driven through innovation related to cross-sectoral issues deriving from the societal demand for changes and related to the tourism sector such as circular economy, biodiversity protection, energy efficiency, use of non-fossil energy sources (solar or wind energy), water saving policies, valorization of the blue economy (sustainable use of oceans, river and groundwater resources together with the protection of coastal lines). All these issues are directly and indirectly connected to tourism and contribute to spread a new model of tourism whose foundations are based on the awareness of the importance of environmental sustainability of both tourist supply and demand (European Parliament, 2025; UN, 2024).

The cross-sectoral nature of possible innovation arising from a new model of tourism and absorbed from the whole economy from the tourism itself makes particularly difficult to measure innovation in the sector. Here a measure to account for the whole sector's ability to produce new impulses to the growth and the repositioning of the tourism sector is proposed (Canale & De Siano, 2025). It is calculated as:

$$\text{TOUR\_INNO} = \frac{\text{GERD}}{\text{GDP}} * \frac{\text{TOUR\_EMPL}}{\text{TOTAL\_EMPL}} \qquad (3.1)$$

where GERD is the general expenditure in research and development, GDP is the gross domestic product, TOUR_EMPL is the percentage of employment in the tourism sector compared to the total employment, TOTAL_EMPL. The strategy behind this indicator depends on the composite nature of tourism goods often closely interrelated with other productive sectors and whose contribution to GDP can only be calculated with indirect estimation methods. Furthermore, innovations and investment in research from other sectors spread the effects in the tourism sector, contributing to its development trajectory. This is an attempt to measure how tourism innovate and contemporaneously how innovation in other sectors spread their effects on tourism sector. The indicator may change both because of changes in research and development expenditure as percentage of GDP and for a higher contribution of tourism to total employment. It is the combined effect of the two measures that is able to register the competitive feature of the tourism industry in respect to the national economy. This indicator provides a holistic approach to the investigation of tourism innovation contribution to the environment. Innovation in the tourism sector is complex in nature and ranges from innovative management of destinations to internal reorganizations of individual companies to public interventions to protect cultural, landscape and environmental assets, not to mention the new technologies that help in their management and fruition. All these factors contribute together to the expansion of tourism in the direction of innovation. Rescaling general expenditure in research and development with the weight of tourism employment in respect of total employment is able to capture both the general attitude of a territory in respect to innovation and the contribution of tourism to the economy in terms of employment. This measure of tourism innovation, despite simple in principle, allows, through a macroeconomic perspective, to account for the complex nature of the sector and the effects the other sectors exert on tourism and vice versa.

Investigating the connection between this measure of tourism innovation and carbon emissions intensity per GDP (CE) in selected European countries (Austria, Belgium, Denmark, Finland, France, Germany, Greece, Ireland, Italy, Netherlands, Norway, Portugal, Spain, Sweden and UK), it is observable that the higher the average degree of innovation in tourism sector is the lower the average carbon emission intensity per GDP (Fig. 3.4) at least in the period 2005–2019.

In tourism, innovative and successful competitors at the product level will soon find themselves being imitated. At the process level innovations tend to be outsourced. Imitation and outsourcing are thus important means of disseminating innovation in the field of tourism. As the focus on delivering sustainable tourism, and the move to a greener tourism

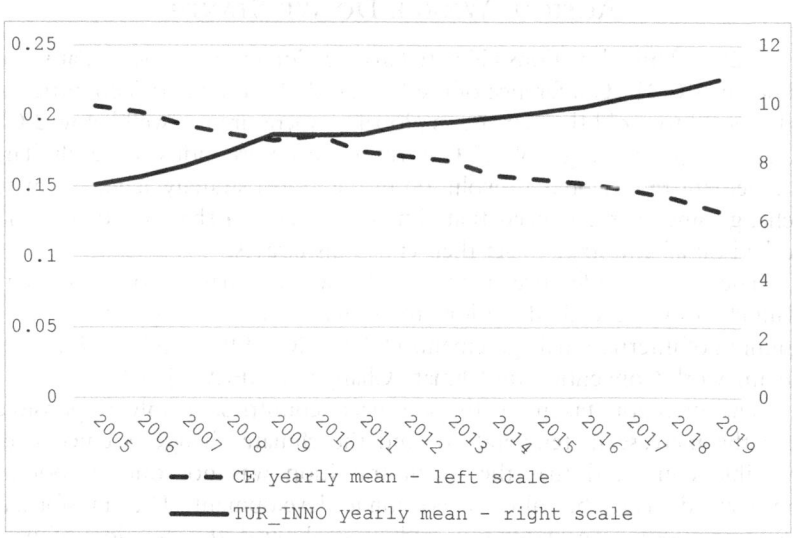

Fig. 3.4 Carbon emission intensity per GDP and tourism innovation in Europe (2005–2019) (*Source* Own elaboration on Eurostat. Data presented in Canale and De Siano [2025])

economy, increases, imitation and outsourcing will to play an important role in the development and adoption of green innovation products and practices by tourism businesses (Canale & De Siano, 2025). It seems that, being tourism a sector highly dependent on environment, the innovation contributed to the reduction of CO_2 emissions paving the way in turn to a more environmentally sustainable tourism.

However, despite innovation in tourism could strengthen the transition towards a greener and sustainable sector and the improvement of environmental quality of local destinations, it should be remembered that the expansion of tourism highly contributes to pollution through the international long-distance trips across continents (Scott & Gössling, 2021). Therefore, at present, the best innovation to introduce in the sector all over the world—waiting for non-polluting flights—is supposed to be the valorization of short distance tourism.

3.5 The Evolution of the Institutional Design About Tourism and Climate Action: Where Do We Stand?

The 2015 United Nations Climate Change Conference, held in Paris also known as COP (Conference of the Parties) 2021, bound 195 countries of the world to avoid the increase in global temperature to "well below 2 C" and aiming for a target of 1.5 C warming above pre-industrial levels. The agreement represented a revolution in the global strategy against climate change since it recognized that climate change is a shared problem and called on all countries to set their emissions targets.

The Paris conference represented the culmination of a strategy launched by the United Nations to combat climate change through the signing of international agreements (UNFCCC, 1997; UNEP, 1987; UN Framework Convention on Climate Change [UNFCCC], 1992).

The direction taken by the advanced countries, mainly responsible for the increase in temperatures and the climate change, seemed irreversible convinced that the green transition was not only a tool to mitigate damage, but also for economic development. The transformation of production systems would have allowed the creation of new growth models for advanced economies that are rather stagnant almost everywhere.

The need to design new strategies for the transformation of the production system has also involved the tourism sector for which international institutions have committed to specifically inflect the path to follow. The coordination was born at the solicitation of the United Nations among the G20 countries in 2010 to enhance the contribution of tourism to the recovery after the economic crisis of 2008. The coordination group of economic policies on tourism within the G20 is known as T20 or Tourism Working Group (TWG) and meets annually to identify strategies. Only in more recent times the focus of the T20 become more cantered on the relationship between tourism and the environment and on the contribution that the sector can provide to environmental sustainability. There has been an acceleration within the G20 TWG held in India in 2023 One Earth, One Family, One Future—Green Tourism, Digitalization, Skills, Tourism MSMEs and Destination Management and in Brazil in 2024—Building a just world and a sustainable planet—Sustainable Tourism, Technical Qualification and Training and Strengthening Tourism Financing as an engine of socio-economic development (for a reconstruction of the meetings see https://www.unwto.org/tourism-g20-economies).

Despite the awareness that tourism had a fundamental role in mitigating the effects of growth on the climate, and the requests that came from the G20 TWG meetings, the COPs meeting, occurred over the years, neglected to address the issue explicitly. The most significant contribution comes from COP26 with the Glasgow Declaration on Climate Action on Tourism (One Planet Sustainable Tourism Programme, 2021). It recognizes the need to accelerate climate action in tourism and to secure strong commitments to support the global goals to halve emissions over the next decade and reach Net Zero emissions as soon as possible before 2050. It is a voluntary commitment which tourism organizations of all kinds subscribe to accelerate and coordinate climate action in tourism. The general objective of reducing emissions is pursued through continuous monitoring of the indicated pathways: measure, decarbonise, regenerate, collaborate and finance.

Public and private institutions, after adhering to the plan, submit to the UNWTO monitoring the progress in implementing the plans, the long-term objectives and the strategies to pursue them. Adherence to the Glasgow Declaration also intends to encourage working in a collaborative spirit, sharing good practices and solutions and disseminating information to encourage additional entities to become signatories and supporting one another to reach targets as quickly as possible.

Thanks to this process started in Glasgow and which allowed the production of a first report (UNWTO, 2024), finally, in 2024 the COP29 held in Baku in Azerbaijan included in the discussion sessions the Thematic Day on Climate Change and Tourism (https://www.unwto.org/news/cop29-to-feature-thematic-day-on-climate-change-and-tourism-for-the-first-time) where the tourism ministers of the participating countries met. The aim of the special session was to gather ever wider consensus around the Glasgow Declaration on Climate Action on Tourism and to monitor through a scientific approach the progress towards the decarbonization of the sector. In Baku, the participating countries formally adopted the Statistical Framework for Measuring the Sustainability of Tourism (MST), suggested by the UN Statistical Commission to uniformly measure at international level the sector's climate action impacts. In this context the Achilles heel is the lack of awareness that climate change related to tourism derives from the ever-increasing number of international flights. The UNFCCC outsourced the international aviation emissions policy management to the International Civil Aviation Organization (ICAO), whose action is significantly circumscribed by its regulatory dictate, and from its inability to coordinate with single nations policy measures about CO_2 emissions (Cames et al., 2023). At the same time domestic issues about air transport emissions are still under the competences of the United Nations, giving rise to conflicts between regulatory measures and compromising the comprehension of the centrality of this issue for the tourism sector. The result is that it is considered little responsible for the emissions caused by the demand for aviation and maritime transports it generates (Lyle, 2024).

It is clear that, at an international level, the awareness that tourism can make a decisive contribution to reducing global warming is growing. There is also a common reference framework to draw inspiration from to promote and preserve the tourism sector with a view to ever greater environmental sustainability. It is felt that the topic is relevant, and that it is desirable to push for the good practices implemented by companies and institutions to be increasingly widespread, focusing on the awareness that global warming can compromise life on earth. However, these are voluntary adherences to strategies, which, despite effective, represent only a strong suggestion rather than a constraint. COP29, the last international meeting on climate held in Baku, achieved the shared goal of tripling financing to developing countries, from the previous goal of USD 100 billion annually, to USD 300 billion annually by 2035, and of securing efforts of all actors to work together to scale up finance to developing countries, from public and private sources to the amount of USD 1.3 trillion per year by 2035. However, no agreement was found on what to do to reduce damage to the environment, but rather it is allowed to foul despite compensations for damages are provided. The result is that, like other economic activities, tourism continues to pollute, especially through the increase in international travel, and that, although it represents a sector in which innovation can contribute effectively towards environmental protection, the result may be modest without radically rethinking the way of doing tourism (Gössling et al., 2024).

REFERENCES

Bruckner, B., Hubacek, K., Shan, Y., Zhong, H., & Feng, K. (2022). Impacts of poverty alleviation on national and global carbon emissions. *Nature Sustainability*, 5(4), 311–20.

Cames, M., Graichen, J., Wissner, N., Faber, J., Nelissen, D., van der Veen, R., Scheelhaase, J., Maertens, S., Grimme, W., Behrends, B., & Jever, M. (2023). *Paris Agreement: Development of measures and activities for climate-friendly aviation and maritime transport* (25/2023). German Environment Agency. https://www.umweltbundesamt.de/sites/default/files/medien/11850/publikationen/25_2023_cc_paris_agreement_0.pdf

Canale, R. R., & De Siano, R. (2025). *Environment, innovation and tourism: Challenges for sustainability in Europe* in Aldieri, L. (2025). *Technological evolution: How innovations are changing our future*. Springer.

European Parliament. (2025). *Circular economy: Definition, importance and benefits*. May 2024 dowloadable at https://www.europarl.europa.eu/topics/en/article/20151201STO05603/circular-economy-definition-importance-and-benefits

Gössling, S. (2002). Global environmental consequences of tourism. *Global Environmental Change*, 12(4), 283–302. https://doi.org/10.1016/S0959-3780(02)00044-4

Gössling, S., & Hall, M. C. (Eds.). (2006). *Tourism and global environmental change. Ecological, social, economic and political interrelationships*. Routledge.

Gössling, S., Humpe, A., & Sun, Y. Y. (2024). On track to net-zero? Large tourism enterprises and climate change. *Tourism Management*, 100, 104842. https://doi.org/10.1016/j.tourman.2023.104842

Gössling, S., & Peeters, P. (2015). Assessing tourism's global environmental impact 1900–2050. *Journal of Sustainable Tourism, 23*(5), 639–659. https://doi.org/10.1080/09669582.2015.1008500

Hubacek, K., Baiocchi, G., Feng, K., & Patwardhan, A. (2017). Poverty eradication in a carbon constrained world. *Nature Communications, 8*(1), 1–9.

International Labour organization (ILO). (2019). *Sustainable tourism—A catalyst for inclusive socio-economic development and poverty reduction in rural areas*. International Labour Organization, Geneva. https://www.ilo.org/sites/default/files/wcmsp5/groups/public/%40ed_dialogue/%40sector/documents/publication/wcms_601066.pdf

IPCC. (2022). In H.-O. Pörtner, D. C. Roberts, M. Tignor, E. S. Poloczanska, K. Mintenbeck, A. Alegría, M. Craig, S. Langsdorf, S. Löschke, V. Möller, A. Okem, & B. Rama (Eds.), *Climate change 2022: Impacts, adaptation and vulnerability. Contribution of working group II to the sixth assessment report of the intergovernmental panel on climate change* (3056 pp.). Cambridge University Press. https://doi.org/10.1017/9781009325844

IPCC. (2023). In Core Writing Team, H. Lee & J. Romero (Eds.), *Climate change 2023: Synthesis report. Contribution of working groups I, II and III to the sixth assessment report of the intergovernmental panel on climate change* (pp. 35–115). IPCC. https://doi.org/10.59327/IPCC/AR6-9789291691647

Lankes, H. P., Macquarie, R., Soubeyran, É., & Stern N. (2024, February). The relationship between climate action and poverty reduction. *The World Bank Research Observer, 39*(1), 1–46. https://doi.org/10.1093/wbro/lkad011

Lyle, C. (2024). *Commentary: Capping aviation emissions—A pressing necessity with a potential solution*. GreenAir. Retrieved June 27, 2024 from https://www.greenairnews.com/?p=5451

OECD. (2013). *Green innovation in tourism services* (OECD Tourism Papers, No. 2013/01). OECD Publishing. https://doi.org/10.1787/5k4bxkt1cjd2-en

One Planet Sustainable Tourism Programme. (2021). *Glasgow declaration: A commitment to a decade of climate action*. https://www.unwto.org/the-glasgow-declaration-on-climate-action-in-tourism

One Planet Sustainable Tourism Programme. (2023). *Glasgow declaration: A commitment to a decade of climate action*. https://www.e-unwto.org/doi/book/10.18111/9789284425242

Peeters, P. M., & Eijgelaar, E. (2014). Tourism's climate mitigation dilemma: Flying between rich and poor countries. *Tourism Management, 40*, 15–26.

Scott, D., & Gössling, S. (2021). Destination net-zero: What does the international energy agency roadmap mean for tourism? *Journal of Sustainable Tourism, 30*(1), 14–31. https://doi.org/10.1080/09669582.2021.1962890

Sen, A. (1979). Utilitarianism and welfarism. *Journal of Philosophy, 76*(9), 463–89.
Sen, A. (2009). *The idea of justice*. Allen Lane.
Smith, K. (2009). *Climate change and radical energy innovation: The policy issues* (TIK working papers on innovation studies, No. 20090101). Oslo.
Torres-Díaz, V., del Río-Rama, M. d. l. C., Álvarez-García, J., et al. (2024). Environmental sustainability and tourism growth: convergence or compensation? *Quality and Quantity*. https://doi.org/10.1007/s11135-024-01906-w
UNEP. (1987). *The Montreal protocol*. https://www.unep.org/ozonaction/who-we-are/about-montreal-protocol
UNFCCC. (1992). *United nations framework convention on climate change*. https://unfccc.int/resource/docs/convkp/conveng.pdf
UNFCCC. (1997). *Kyoto protocol to the united nations framework convention on climate change*. https://unfccc.int/sites/default/files/resource/docs/cop3/l07a01.pdf?download
United Nations. (2024, January 22). *What does 'biodiversity' mean and why is it important?* https://news.un.org/en/story/2024/01/1145772
UNWTO. (2024). *The Glasgow declaration on climate action in tourism*. UNWTO. Retrieved February 5, 2025 from https://www.unwto.org/the-glasgow-declaration-on-climate-action-in-tourism
World Tourism Organization (WTO). (2024). *Glasgow declaration implementation report 2023—Advancing climate action*. UNWTO, Madrid. https://doi.org/10.18111/9789284425242
World Tourism Organization and International Transport Forum (UNWTO and ITF). (2019). *Transport-related CO2 emissions of the tourism sector—Modelling results*. UNWTO, Madrid. https://doi.org/10.18111/9789284416660

CHAPTER 4

Tourism and Sustainability of Destinations: Some Analytical Tools

Abstract This chapter deals with the relationship between tourism and destinations sustainability from a theoretical point of view. The nature of "public goods" of both the environment and tourism resources leads to the use of microeconomic tools to identify the connection between economic choices and the impact on society. A complex picture emerges in which tourism is sustainable if both aspects related to the environment and those related to the balance in society are considered. The natural consequence of this complexity is that the "optimal solutions" depend on the context and the nature of destination to which they are applied and cannot be qualified as universally valid.

Keywords Tourism sustainability · Destinations · Analytical tools · Complexity · Resources · Collective preferences

4.1 The Complexity of the Phenomenon of Tourism Sustainability

The concept of sustainability—as it has been widely portrayed in previous pages—is very complex. It involves research and identification of what should be defined as sustainable, for whom it should be sustainable and at what level it should be sustainable. The investigation is even more complex when we refer to the tourism sector. The evaluation of the relationship between tourism and sustainability implies, on the one hand, a measurement of the benefits and costs in both economic and social terms of the exploitation of tourist resources in the present and, on the other, the possibility that these resources are available over time so that future generations can benefit from it. In more strictly economic terms, tourism sustainability deals with on the one hand with the objective of profitability of the companies involved, with the level of employment and with the social costs connected to the exploitation of tourist resources and, on the other, with long-term sustainability term of the development of the sector, both in relation to the sector itself and for the sectors that are related to it. There are therefore two key terms: profitability—understood not only in an economic sense—and sustainability.

Regarding the sustainability of environmental resources, the *World Commission on Environment and Development* in the report *Our Common Future* defined the conditions for sustainable development as a whole. Sustainable development is "development that meets the needs of the present without compromising the ability of future generations to meet their needs" (World Commission on Environment and Development, 1987. For a review of the literature relating to sustainable development see Mensah, 2019). It is therefore felt that there is a *trade-off* between present use and future use which, in the case of tourist resources, will concern the possibility of using environmental, landscape and archaeological resources today, without compromising their availability for the future. This objective must then be reconciled with the ability to produce income and ensure a good quality of life.

With the evolution of the literature related to tourism and its sustainability, analytical models have moved from the analysis of the life cycle of the tourist product in a strictly economic sense (Butler, 1980), to the analysis of the change in the relationship between tourists and the territory (Liszewski, 1995), to the environmental effects of the development of the sector (Zaręba, 2010). This literature, which reconstructs the evolution in both qualitative and quantitative terms of the tourism product over time, allows us to have a reference framework from which to draw inspiration to ensure that the sector does not evolve in a way that does not compromise its profitability, the relationship with the territory and the surrounding environment (for the evolution of the literature see Butler, 1999; Butowski, 2012).

With a simplification, it is possible to state that the development of the tourism sector can be considered sustainable when there are:

1. Use of tourist resources not exceeding their ability to regenerate
2. Maintaining the flow of waste in the environment not exceeding the degree of assimilation capacity.

These two objectives must be reconciled with the capacity of the tourism system to

3. Produce income and guarantee an adequate standard of living for the local population.

The standard of living is not only related to the achievement of economic objectives, but also to a series of general conditions of the economic system of the place where the tourist resources are located. They may include the level of education or democratic participation as well as the quality of life of residents which could be threatened by excessive congestion of the area caused by the presence of tourists.

The total tourist resources available to a territory are represented—in a simplified way—by the sum of the man-made resources with the environmental and cultural resources (Tenuta, 2009).

$$R_T = R_U + R_{N,H} \qquad (4.1)$$

where R_T represents the total tourist resources, R_U the man-made resources and $R_{N,H}$ the resources coming from nature (N) or cultural heritage (H). The man-made economic resources are potentially infinitely reproducible, but their growth can compromise the existence of tourism resources in the strict sense. Tourism sustainability is guaranteed if the growth of tourism resources is not compromised over time. However, it is possible to refer to two ways of understanding this sustainability.

The first is sustainability in strong form. Tourism sustainability is meant in a strong form if the following condition holds:

$$\Delta R_T \geq 0$$

and, at the same time,

$$\Delta R_U \geq 0 \cap \Delta R_{N,H} \geq 0 \qquad (4.1.1)$$

That is, if the growth of human-produced resources does not compromise the current state or the growth of natural or cultural resources. The "intersection" symbol ∩ indicates that the two conditions must be valid simultaneously, i.e. that the first without the second condition does not describe states of sustainability of the tourism sector growth.

However, it is meant in a weak form if to the condition:

$$\Delta R_T \geq 0$$

and contemporaneously:

$$\Delta(R_U + R_{N,H}) \geq 0, \quad (4.1.2)$$

i.e. if the sum of growth of man-made resources plus environmental and cultural resources exceeds or equals zero. In other words, in this second vision tourism is said to be sustainable if the growth of resources produced by man at least compensates for the consumption of environmental and cultural resources that follows the excess growth of economic activity (and vice versa).

It is clear that in the first case the development of a territory is achieved without damaging environmental and cultural resources, while in the second case there is the willingness to give up a part of the tourist resources endowment in order to guarantee the economic expansion of the territory. These two representations describe in a simplified form the possible development paradigms to draw inspiration from. These paradigms, however, can vary depending on the state of development of a country and the perception that residents have of the congestion resulting from the implementation of tourist activity. It is therefore very difficult to account for all the variables that define the possible sustainable tourism development strategies.

The evaluation of tourism sustainability is a very complex concept that involves the conciliation of multiple aspects that often come into conflict with each other and which it is necessary to make explicit in order to be able to evaluate the benefits and costs of the tourism development of a territory.

As stated in the European Charter for Sustainable Tourism—deriving from the general principles defined in Agenda 21 in Rio de Janeiro in 1992—sustainable tourism represents *"any form of development, organization or tourist activity that respects and preserves natural, cultural and social resources in the long term, and contributes in a positive and equitable way to economic development and to the improvement of the quality of life of people who live, work or stay in protected areas"* (Europark Federation, 2010, p. 4).

Figure 4.1 schematically presents the complexity of the sustainable tourism phenomenon, summarizing the problems that need to be reconciled. The diagram shows that to have sustainable tourism it is necessary to account for three types of objectives: economic objectives, social objectives and conservation objectives. These three objectives often come into conflict with each other to the point of compromising the full realization of all three at the same time in the present and making it difficult to maintain their balance over time.

On the left of the diagram the economic objectives are indicated: among the main ones are growth and income distribution. Growth represents the rate of increase in GDP, which in modern market economies is the primary indicator for evaluating the expansion capacity of an economic system; however, this must also be associated with an evaluation of how these additional resources produced are distributed. An economic objective must consider the distribution of income: an excessively unequal distribution exacerbates social conflict and calls into question the very premises for production. On the opposite front, an excessively egalitarian distribution, according to some scholars, reduces the incentives for economic progress. Finally, efficiency, i.e. the ability of a production system to achieve the best possible results with the available production factors, is considered as further economic objective. However, if it is simple to identify a condition of greater efficiency when a single good is produced—production of the same quantity of product with a smaller quantity of inputs or a greater product with the same production factors—it is much more complex to establish a single efficient result

Fig. 4.1 The complexity of the objective of sustainable tourism (*Source* Own elaboration)

when production factors must be combined to produce two alternative products. In this case we are talking about an efficiency frontier—that is, a series of optimal combinations of two outcomes, such as tourist employment and environmental protection—without however knowing a priori which is the most "convenient" from the social point of view.

In the centre, the so-called social objectives are indicated, namely social cohesion, cultural identity and mobility. A territory with a tourist vocation cannot fail to account for the congestion caused by the excessive presence of tourists and the conflict between residents and operators. Likewise, tourism management that effectively uses local resources guarantees greater social cohesion and openness to alternative scenarios of societal development. Similar considerations can be made regarding cultural identity: a local community traditionally open to exchanges will have a greater aptitude for welcoming tourists, accustomed to building its own identity also through contamination between cultures. The opposite happens in communities less accustomed to interactions, which will see the excessive presence of visitors as a threat. Finally, the social objectives include the mobility of residents and non-residents. It is obvious that the excessive presence of tourists can represent a factor of excessive congestion in the area capable of compromising both economic and social cohesion objectives.

Finally, on the right, the conservation objectives of tourist resources are indicated. Returning to the concept of sustainability, it is obvious that the tourist supply can be kept constant over time only if the cultural assets and/or the environment that define it are preserved. This implies that the objective of maximizing economic results today may call into question the achievement of the same tomorrow. They are also related to the management of the territory and its recovery capacity—the so-called "resilience"—the waste management cycle and the environmental regeneration that the territory and local institutions are able to guarantee.

The contents of the three objectives often overlap, highlighting the complexity of the sustainable management of tourist resources: it is complex because the definition of each objective depends on the value that each local system assigns to it today, also in anticipation of what will happen tomorrow.

4.2 The Evolution of the Tourism Product and the Life Cycle Approach from the Perspective of Sustainability

Tourism resources have long been the subject of investigation in economic literature. A brief reference to the founding contribution from which the subsequent literature developed is useful for deepening the relationship between tourism and sustainability. The model presented below constitutes a benchmark to refer to address the complex relationship between tourism and sustainability in light of the different phases of its evolution.

The Canadian geographer Butler (Butler, 1980), using the theoretical framework of the product life cycle developed for consumer goods (Katona, 1964; Vernon, 1966), described the evolutionary trajectory to which a tourist destination is subject, and its possible reconversion paths once it has exhausted its cycle. This trajectory—which can also be read from the perspective of sustainability—can be summarized in the following phases represented with a straightforward image in Fig. 4.2.

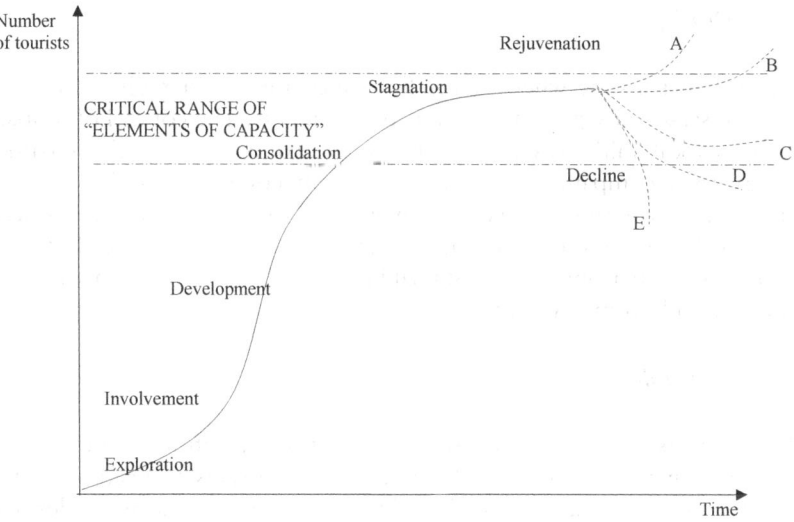

Fig. 4.2 Tourism Area Life Cycle Model (*Source* Butler [1980, p. 7])

1. Exploration

A small number of guests "discover" a destination. It presents an uncontaminated character and an interpenetration with local life. At this stage there are few tourists, and they can enjoy the location without coming into conflict with the local community and without compromising environmental sustainability.

2. Involvement

As the location begins to gain notoriety, the number of tourists increases. Residents contribute to the construction of a more complex range of services. The number of tourist establishments increases and part of the local facilities are converted to accommodate an ever-increasing number of guests. The interaction between residents and tourists in this phase is still virtuous, since this phase of expansion is also accompanied by the construction of infrastructures that improve both the quality of the tourist supply, but above all the quality of life of the residents.

3. Development

Then comes a time when tourist flows increase considerably, becoming, in high seasonality periods, greater than residents. The control of tourism escapes local initiatives and increases the exploitation of the territory by external companies. The latter move in conflict with the territory to continue offering services, the quality of which, however, is reduced due to the excess use of the structures and the overcrowding. Both the problem of environmental sustainability and the conflict with the local population become evident.

4. Consolidation

It represents the terminal phase of the positive evolution of the destination, in which the growth in the number of tourists is reduced over time (decreasing marginal trends). The location is one of the main destinations of the tourism industry where all the major international hotels are represented, but there is no more room for further progress in the sector.

5. Stagnation

At this stage the maximum number of tourists has been reached and the destination is no longer attractive. The number of tourists no longer increases over time (constant marginal trends). The location relies on repeat trips from habitual travellers; considerable efforts are required to keep the number of visitors constant. Environmental, social and economic controversial issues arise. At this stage two paths of subsequent evolution can emerge, depending on the strategy followed by the institutions and local actors.

6. Decline

Environmental degradation and conflict with the territory make the location no longer attractive. The number of tourists reduces, as well as of the accommodation facilities and the quality of the services offered (trajectories D and E in Fig. 4.2). To survive, the tourist location must focus on alternative strategies that rebuild its image starting from new premises.

7. Rejuvenation

These new premises represent the new bases from which to start again. Local actors, as well as institutions, contribute together to re-establish or recreate a different image of the destination by focusing on a different tourism that resolves the conflict with the territory, the environment and the local population (trajectories A and B in Fig. 4.2).

In Fig. 4.2 it is possible to observe a wide range of possibilities for the evolution of the tourist destination after the stagnation phase. Once the tourist product has entered the critical range of the "elements of capacity"—or the carrying capacity of the territory—the path of its evolution will depend on the strategies adopted and on the implementation of virtuous or conflictual strategies between the institutions, the choices of intervention and local actors.

This model only provides a stylized representation of the evolution path of the tourism destination, but it helps us to understand how the topic of tourism sustainability could be expressed according to its evolution through time. It therefore helps to understand that, in order to encourage positive evolution, it is necessary to review in time the foundations on which the destination itself was built, to avoid its definitive decline.

4.3 Sustainable Tourism and Market Equilibrium: The Effects on the Quality of Life

Before proceeding to an analytical representation of the concepts involving sustainable tourism, it is appropriate to recall these concepts from a different perspective to bring out the ways in which conflicts between the exploitation of tourist resources and their conservation emerge. The first conceptual area concerns the complexity of the relationship between economic activity and the environment. In fact, for the production system, environmental protection expenses represent a major cost factor for firms and local residents seeing resources subtracted from their current income. However, the environment and cultural resources offer numerous economic opportunities which, to be fully exploited, require a medium-long-term strategic and cognitive horizon and a considerable availability of human and financial resources. Economic activities producing income in many cases release effects on the environment, in the same way if the environment is not protected and preserved, the ability to produce income and improve the quality of life is compromised. The second conceptual area—belonging to a holistic approach—concerns the fact that tourist resources must be examined in a broad sense, i.e. taking into consideration all the aspects that involve them: nature, cultural resources, social aspects, etc. Resources have different characteristics, and their use could generate conflicts with the local community as well as distribution conflicts if everyone is not guaranteed adequate remuneration for the resources used and the pressure suffered. The third conceptual area concerns the endogenous relationship between tourism understood as an economic activity and the environment. Tourist activities use the environment and the supply of cultural resources as a productive resource. The existence of natural and cultural resources is therefore a prerequisite for the birth of tourism activities. The latter, at the same time, are able to affect the natural and cultural environment in which they are implemented through pollution or congestion (for a rigorous theoretical framework and for an empirical analysis of the relationship between the environment and growth of tourism activity see Pulido-Fernandez et al., 2019). Finally, all these alternatives can be declined by referring to some analytical simplifications that make more clearly the way in which economic aspects and the issue of resource protection intersect in defining tourism sustainability. The analytical tools of economic theory are useful for representing the complexity of the topic of tourism sustainability. Of course, they are not able to account for the multiple facets to which the phenomenon lends itself, but they can help to understand why it is not sufficient to refer to the simple rules of competition to analyse supply and demand of tourist services. This circumstance depends on the fact that tourist production uses and affects the availability of production factors which have the characteristic of a public good and which, therefore, do not find an adequate measure in the simple rules of supply and demand.

Through Fig. 4.3 it is possible to describe this circumstance. The price of a tourist service in a specific place is represented on the x-axis, while the quantity of tourist services is represented on the y-axis. The continuous decreasing curve D_T represents the relationship between price and quantity on the demand side. As microeconomics suggests to us, it is the result of the consumer's optimal choice and reveals the average revenue each producer obtains for each unit sold. In the presence of "normal" goods, it is represented by a decreasing relationship: the higher the price of the tourist service, the lower the quantity demanded. The continuous S_T curve represents the supply of tourism services. As is known, it represents the aggregate marginal cost curve above the average cost curve. These costs are those that companies can directly measure, and which enter into their financial statements. In a short-term horizon, all costs relating to inequality, damage to health caused by pollution and the erosion of environmental heritage which will make tourism less sustainable in the future will be excluded from the measurement. In other words, the negative externalities linked to the production of tourist services will not be considered.

Therefore, what the market equilibrium is able to return is the price that equals the quantity demanded to the quantity supplied without taking into account the sustainability of the tourist supply or the implications on the society in which the activity is carried out. This price is represented by the value P_0 to which the equilibrium quantity Q_0 corresponds.

If, however, the social costs of the production of tourist services is taken into account, the supply function shifts to the left in S_T' since an additional value generated by the social costs resulting from the increase in tourist activity should be added to each directly measurable cost. It

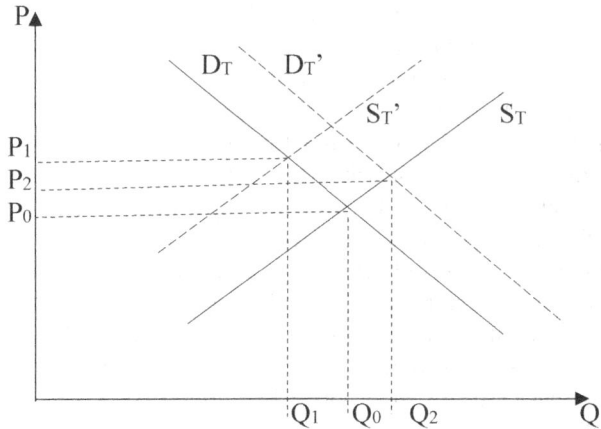

Fig. 4.3 Social equilibrium and market equilibrium in the tourism sector

is clear that, in this second case, the market equilibrium price would be higher ($P_1 > P_0$) and the equilibrium quantity of tourist services would be lower ($Q_1 < Q_0$). Alternatively, it could also be considered the circumstance that the presence of a greater availability of tourist services creates positive externalities on the supply side (reduction of costs for tourism businesses operating in the same district) which would translate into an equilibrium generated by a greater quantity and a lower price.

Even the demand for tourist services may not be correctly detected by the market: for example, the presence of a high number of visitors interested in the cultural aspects of the area could generate an image of quality making all travellers willing—for the same conditions—to pay a higher price. This circumstance would be described by an equilibrium price equal to P_2 and an equilibrium quantity of tourist services equal to Q_2.

The same phenomenon can also be described from a second point of view, i.e. by relating the number of tourism businesses to the product of each of them. In Fig. 4.4 the number of tourist activities (A_T) is represented on the x-coordinates, while the tourist product (P_T) or the services offered are represented on the y-coordinates. The uppermost descending curve RM_e represents the average return of each tourist activity as the number of companies increases: it is decreasing because each company will have a lower profit on average if there are many firms on the market offering the same service. If, for simplicity, we assume that the cost of an activity is constant (C_U curve), the market equilibrium will be given by the matching between the cost curve and the average yield curve corresponding to a number of tourism businesses equal to AT_0. However, there is the marginal social return of tourism businesses, which indicates the increase in return for the community following the entry of one more unit on the market. As is known, the RMs curve that represents it lies below the average revenue curve, both for reasons linked to the market (each additional company reduces the performance of the others) and for reasons linked to the effects on the community (congestion, pollution, etc.) of the increase in tourist activity. The efficient equilibrium would be in correspondence with a number of firms AT_1 lower than AT_0. As in the previous case, the market is not able to restore a balance accounting the social costs and benefits of tourism.

This description of the inefficiency of the market in achieving a balance that guarantees the sustainability of tourism can also be represented in terms of the relationship between quality of life and use of tourist resources depending on the different phases of development of a territory and the strategy that the institutions they intend to pursue. Figure 4.5 presents the possible relationships between use of tourist resources and increase in quality of life (Pearce & Turner, 1989).

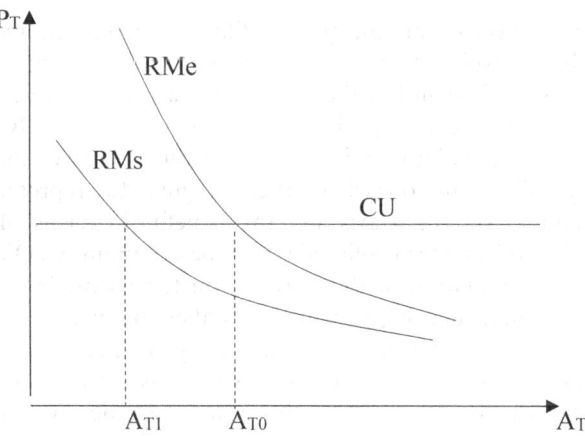

Fig. 4.4 Common resources and the output of tourism firms

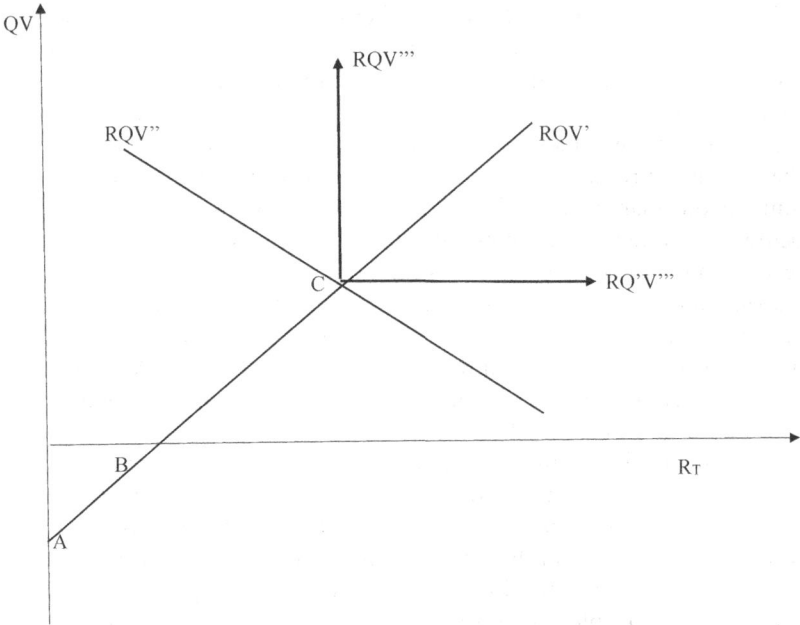

Fig. 4.5 Exploitation of tourist resources and quality of life

The objective is to relate the quality of life, both in economic and social terms, with the exploitation of tourist resources to observe how in the different phases of expansion the two values can come into conflict. The type of conflict between objectives will be defined differently depending on the reference paradigm and the value of the objectives that the local community will assign to each of them. Figure 4.5 represents the R_T tourist resources on the x-axis and the benefits in terms of improved quality of life and greater profitability for the community (QV) resulting from greater exploitation of the territory on the y-axis. It is possible to draw a first straight line RQV' which describes the growing relationship between use of tourist resources and quality of life. As can be imagined, the positive slope of the line shows that, as the exploitation of R_T increases, the quality of life increases because incomes and employment increase too. Furthermore, since the level of development is very low, the perception of the damage created by the reduction of the availability of resources is very limited. We can assume that at the beginning the straight line crosses the ordinate axis in correspondence with negative values of quality of life (point A) and that it cuts the x-axis in correspondence with the positive value of tourist resources (point B). In other words, for very low values of resource use, living standards go from negative to positive and then proceed following a positive connection. In this case, the sustainability of the tourist activity is designed in such a way as to combine the increase in the well-being of the community with an ever-increasing rise in the exploitation of resources. The concept of sustainability is unbalanced in favour of economic results, since it is believed that an ever-increasing exploitation of resources is not only possible, but always capable of generating increasing benefits for the community.

On the contrary, the RQV" line describes an inverse relationship between resources and quality of life. In other words, following this path of development, a greater quality of life can only be achieved by reducing the exploitation of tourist resources. The issue of sustainability in this case is linked to the idea that better economic and social results can only be achieved with a policy of recovery and less exploitation of the resources that the environment and history have made available to human-being. On the other hand, greater exploitation of resources would lead to a worsening of the quality of life and a reduction in benefits for the community.

The line RQV''' describes, again, a path of expansion in which the increase in the quality of life does not depend on the greater exploitation of tourist resources, which, being used to the maximum, can no longer produce additional benefits for the community. Finally, at least in principle, it is possible to represent through the horizontal line RQV'''' a situation in which, despite being increasingly exploited, tourist resources are not able to affect the quality of life, which instead remains constant.

4.4 Tourism Sustainability and Collective Preferences

The phenomenon of sustainability can also be examined from a further point of view to indicate how the comparison between benefits and costs intersects with the preferences of firm and residents in relation to the trade-off between the use of tourist resources and the size of the tourist activity (Johnston & Tyrrell, 2005).

This relationship depends on three functions: the first one which sees, following the most accredited line of reasoning, the benefits for the community to grow as the use of tourism resources increases. In analytical terms:

$$B_T = f(R_T) \tag{4.2}$$

with

$$f'(R_T) = \frac{\Delta B_T}{\Delta R_T} > 0 \text{ e } f''(R_T) = \frac{\Delta B_T}{\Delta R_T} < 0 \tag{4.2.1}$$

That is, the benefits grow as the exploitation of tourist resources increases, but they grow in a decreasing manner (the function is concave downwards).

The second function is represented by the costs for the community which, as in the case of benefits, increase as the use of tourism resources increases. The following analytical definition then applies:

$$C_T = g(R_T) \tag{4.3}$$

with

$$g'(R_T) = \frac{\Delta C_T}{\Delta R_T} > 0 \text{ e } g''(R_T) = \frac{\Delta C_T}{\Delta R_T} > 0 \tag{4.3.1}$$

Unlike the case of benefits, the second derivative is greater than zero (the cost function of the use of tourist resources is convex), indicating—following the sustainability paradigm—that costs grow increasingly as the use of tourist resources increases.

These two relationships are depicted at the top of Fig. 4.6. Following the hypotheses formulated in the equations indicated above, the B_T tourism benefits curve follows an increasing trend. The incremental relationship between benefits and use of tourism resources gradually reduces until it approaches zero. The C_T curve also expresses an increasing relationship between costs and use of resources. However, in this case the incremental ratio grows because it captures the hypothesis that as the use of resources increases, the community is affected in an increasingly accentuated manner due to the resulting effects of saturation of the territory. These hypotheses on the performance of the benefit and cost functions can be changed depending on how the concept of sustainable tourism is articulated in the representation. For example, we could assume linear or constant benefits, as well as exponentially increasing social costs, etc.

4 TOURISM AND SUSTAINABILITY OF DESTINATIONS: SOME ... 71

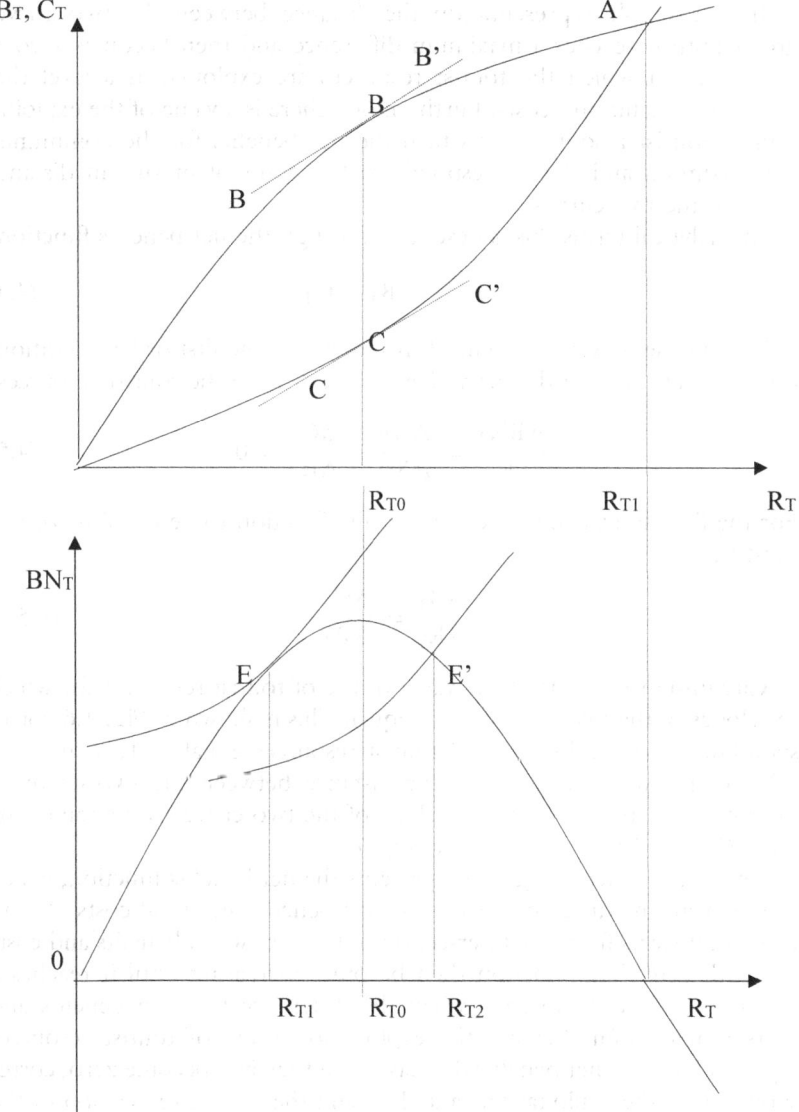

Fig. 4.6 Maximization of the net benefits of tourism and preferences of firms and residents

In our simple representation the distance between the two curves grows until it reaches a maximum difference and then becomes zero at a point A, at which the tourist resources are exploited at a level that equalizes benefits and costs. Furthermore, there is a value of the exploitation of tourist resources for which the net benefits for the community are maximum and this corresponds to the point of maximum distance between the two curves.

In analytical terms this is described through the net benefits function:

$$BN_T = B_T - C_T \qquad (4.4)$$

To find the maximum point of this function (the first order condition) it is necessary to take the derivative with respect to the tourist resources.

$$\frac{\Delta BN_T}{\Delta R_T} = \frac{\Delta B_T}{\Delta R_T} - \frac{\Delta C_T}{\Delta R_T} = 0 \qquad (4.5)$$

For the first derivative of the net benefits function to be equal to zero it must be

$$\frac{\Delta B_T}{\Delta R_T} = \frac{\Delta C_T}{\Delta R_T} \qquad (4.5.1)$$

Circumstance occurs at the value of use of tourist resources for which the slopes of the two functions are equal. This is shown in Fig. 4.6 corresponding to an exploitation of tourist resources equal to R_{T0}. For this value of resource exploitation, the distance between the two curves is maximum (section BC) and the slope of the two curves is the same (the lines BB' and CC' have the same slope).

The lower panel of Fig. 4.6 represents the net benefits function, which results from the difference between total benefits and total costs. As can be seen, it starts from the intersection of the axes when benefits and costs are both zero. The function then becomes increasing until it reaches a maximum point at R_{T0}, i.e. when the difference between benefits and costs is maximum. Beyond the exploitation value of tourist resources equal to R_{T0}, the net benefits decrease until they first become zero, corresponding to the exploitation value R_{T1} and then negative (section of the curve below the x-axis).

The bell-shaped curve of net benefits can also be interpreted as a sort of constraint to which a tourism system is subjected if it wants to obtain at least positive net benefits for the territory on which it is located. In fact, given the curves of total benefits and total costs, all the points above are not achievable with the present technologies or organizational methods, without calling into question the value of the sustainability of tourist services, while all the points below are not efficient. The bell curve therefore represents a sort of trajectory that local and national economic policy authorities should follow in supporting and enabling the expansion of the tourism sector.

On the same graph it is possible to represent the indifference curves that describe how the local community is willing to combine the net benefits of the tourism sector with the exploitation of resources in order to keep the quality of life at least constant. These indifference curves have a positive slope since they express the combinations between a good (the net benefits of tourism) and a bad (the exploitation of tourist resources). They are concave upwards because it is necessary—as the exploitation of tourist resources grows—to make the net benefits of the activity grow more than proportionally. Higher levels of quality of life are achieved by moving from right to left or—with the same exploitation of resources—with greater net benefits. The optimal solution is reached at the point where the indifference curve equals the slope of the bell curve of net benefits.

In analytical terms:

$$\frac{\Delta BN_T}{\Delta R_T} = \frac{U'BN_T}{U'R_T} \qquad (4.6)$$

This equilibrium condition is the one identified on the graph, in the lower box of Fig. 4.6, at point E.

This point E corresponds to the exploitation of resources that maximizes the quality of life of citizens, therefore both individuals and firms. But these preferences could be different depending on the social groups to which they refer. For example, tourism firms could be interested in a greater exploitation of tourism resources for the same net benefits achieved, while individual citizens could prefer to achieve a higher level of net benefits with a lower exploitation of tourism resources. In this case, the indifference curve on the left in the graph (bottom of the Fig. 4.6) can be interpreted as representing the preferences of individuals, while the lower one as the alternatives that generate the same quality of life for firms. When comparing them with the bell-shaped curve, an optimal solution in the Pareto sense is achievable only for private individuals, while for firms the choice is suboptimal. In fact, the solution is represented by the intersection of the two curves at the point E' where the preferences and the constraint coincide, but it would be possible to increase the well-being of the community by reducing the exploitation of tourist resources, i.e. by moving from R_{T2} to R_{T1}.

In other words, the two choices are both possible, given the sustainability constraint, but it is possible that the short-sightedness of some and the difficulty in making preferences explicit generate solutions that are not the most convenient for the community.

It is still possible to hypothesize different shapes of the indifference curves (Fig. 4.7). For example, if it is believed that the exploitation of tourist resources is not an evil, but rather a good that should be exchanged with the net benefits for the local community, the indifference curves will be negatively sloped and the optimal solution for the community will be in the right section of the bell-shaped curve which corresponds to a high exploitation of resources and rather low net benefits (balance point in E_2 with exploitation of tourist resources R_{T2} and net benefits BN_{T2}). Again, if the indifference curves are vertical or the net benefits of tourist services represent a good to which the local community is indifferent, only two optimal solutions will exist: the first with zero exploitation of tourist resources and zero net benefits and the other with maximum exploitation and zero benefits. This second solution is indicated at the equilibrium point in E_3 with exploitation of tourist resources R_{T3} and net benefits equal to zero. Finally, if the indifference curves are horizontal - i.e. tourist resources are a good in which the community is not interested—the optimal solution can be identified at its maximum point of the bell curve: equilibrium point in E_0 with exploitation of tourist resources R_{T0} and net benefits BN_{T0}). This conclusion is somewhat paradoxical since precisely when the tourist resource is indifferent to the community it is possible to derive the maximum net benefit from it.

4.5 Concluding Remarks

This chapter provided analytical tools to analyse the sustainability of tourism. As in the case of any complex discipline dealing with society, it is not possible to have certain answers, but only to know the elements to account for arriving at a choice. The discussion up to this point has shown that the evaluation of tourism sustainability relies on very complex variables. Firstly, it would be necessary to define the costs and secondly

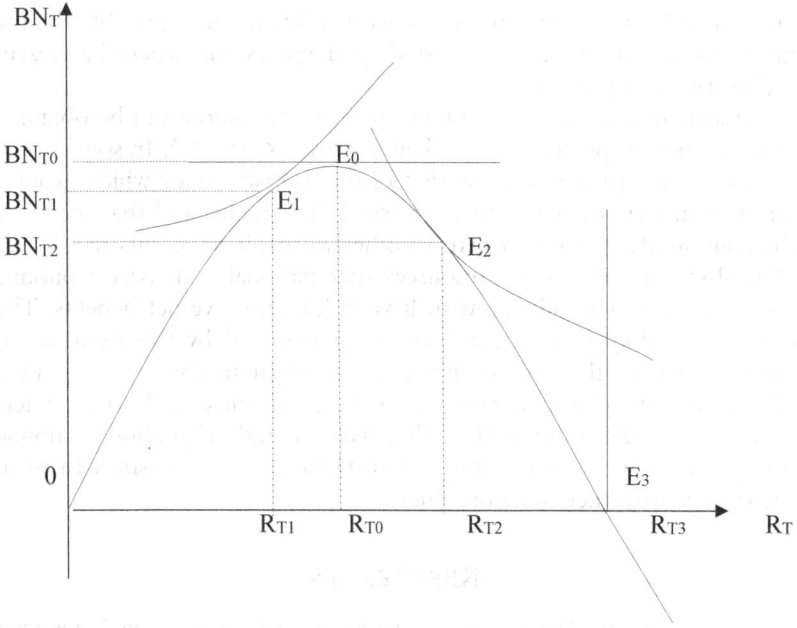

Fig. 4.7 Net benefits and alternative indifference curves

it would be necessary to define the benefits for the community. Both the costs and the benefits are not only those strictly linked to the tourist economic activity but also those linked to the social impacts on the territory both in the short and long term. What perspective do you adopt? The immediate one or the long-term one? Wanting to stick to the sustainability paradigm, it is necessary to take into account that the exploitation of today's resources must not compromise the possibility that future generations will also benefit from the same resources. The question then becomes, how to include future sustainability in the benefit and cost functions? If the future is also included in these functions, do economic policy authorities have sufficient information to predict today what will happen tomorrow? Thirdly, to guarantee appropriate solutions, it is necessary to know the preferences of the community and ensure that the exploitation of resources is precisely that, coupled with a given net benefit value, guarantees the best possible quality of life. How are individuals living in the present able to incorporate the satisfaction of future generations into their preferences? As the present experience teaches us, it is possible for the future to be incorporated into today's preferences only when the present is also truly compromised. An analytical framework on how and what to choose can be obtained from the bell-shaped curve described in Figs. 4.6 and 4.7. In some way it represents the frontier of expansion of the tourism sector which guarantees that at least the costs do not exceed the benefits and that therefore the community is not damaged. In other words, it represents the degree of exploitation of tourism resources that national and local economic policy authorities should allow to have at least positive net benefits. The optimal solution, however, will not be guaranteed by this exploitation frontier, but by the weight that citizens assign to the two objectives of exploitation of tourist resources and net benefits, or by the preferences of the local communities. They have the task of guiding economic policy action and guaranteeing the sustainability of tourism activity in accordance with their scale of values.

References

Butler, R. W. (1980). The concept of a tourist area cycle of evolution: Implication for the management of resources. *The Canadian Geographer, 25*, 151–170.
Butler, R. W. (1999). Sustainable tourism: A state-of-the-art review. *Tourism Geographies, 1*(1), 7–25.
Butowski, L. (2012, April 18). *Sustainable tourism—A model approach. Visions for Global Tourism Industry—Creating and Sustaining Competitive Strategies.* Murat Kasimoglu, IntechOpen. https://doi.org/10.5772/37718. https://www.intechopen.com/books/visions-for-global-tourism-industry-creating-and-sustaining-competitive-strategies/sustainable-tourism-a-modeling-app roach
Europark Federation. (2010). *European charter for sustainable tourism in protected areas.* Regensburg, Germania, EUROPARC Consulting Limited. http://www.parks.it/federparchi/PDF/la_carta.pdf. https://www.europarc.org/library/europarc-events-and-programmes/european-charter-for-sustai nable-tourism/
Johnston, R. J., & Tyrrell, T. J. (2005, November). A dynamic model of sustainable tourism. *Journal of Travel Research, 44*, 124–134. https://doi.org/10.1177/0047287505278987
Katona, G. (1964). *Analisi psicologica del comportamento economico.* Etas Kompass, Milano.
Liszewski, S. (1995). *Przestrzeń turystyczna, Turyzm, 5, 2.* Uniwersytet Łódzki.

Mensah, J. (2019). Sustainable development: Meaning, history, principles, pillars, and implications for human action: Literature review. *Cogent Social Sciences*, *5*, 1653531. https://doi.org/10.1080/23311886.2019.1653531

Pearce, D. W., & Turner, R. K. (1989). *Economia delle risorse naturali e dell'ambiente*. Bologna, Il Mulino.

Pulido-Fernandez, J. I., Cardenas-García, P. J., & Espinosa-Pulido, J. A. (2019). Does environmental sustainability contribute to tourism growth? Ananalysis at the country level. *Journal of Cleaner Production*, *213*, 309–319. https://doi.org/10.1016/j.jclepro.2018.12.151

Tenuta, P. (2009). *Indici e modelli di sostenibilità*. Milano, Franco Angeli.

Vernon, R. (1966). International investment and international trade in the product cycle. *Quarterly Journal of Economics, LXXX*(2).

World Commission on Environment and Development. (1987). *Our common future*. Oxford University Press.

Zaręba, D. (2010). *Ekoturystyka*. Wydawnictwo Naukowe PWN.

CHAPTER 5

Contrasting Impacts of Tourism Expansion

Abstract Tourism growth and its peculiar features are source of multiple phenomena having both negative and positive impacts on the territory and the economy. It can be a cause of excessive congestion and at same time instrument of rapid recovery after a shock, because of its resilient nature. This chapter provides analytical tools to understand and measure the phenomenon of overtourism together with a theoretical description of the broad concept of economic resilience and its application to the tourism sector. The aim is to provide a framework to understand the trade-offs between the ability to contribute to recovery and growth on one side and the congestion of the territory on the other side due to the tourism sector expansion.

Keywords Overtourism · Indicators · Destination management · Resilience

5.1 The Controversial Effect of Tourism on the Territory: The Phenomenon of Overtourism

The complexity highlighted in the previous pages of the relationship between tourism, environment and society both in the present and in the future brings us to identify analytical categories to measure the controversial impacts of tourism. Literature confirms the presence of positive effects of tourism on economic growth mainly attributable to income and employment opportunities, investment in infrastructures and improved balance of payments (Balaguer & Cantavella-Jordà, 2002; Dritsakis, 2012; Durbarry, 2004; Fahimi et al., 2018; Santamaria & Filis, 2019).

Alongside these contributions, however, there is also evidence that an excessive growth of tourism may generate unfavourable results (Albaladejo & Gonzalez-Martínez, 2019; Ehigiamusoe, 2020; Poa & Huang, 2008). The literature identifies these unfavourable results with issue of overtourism, a neologism referring to a phenomenon already known in literature linked to the negative consequences generated by a continuous and indiscriminate growth of tourist flows (Butler, 1980). There is no univocal and objective definition of overtourism since its declination changes depending on the nature of the destination, the space and the time in which it is measured. Overtourism mainly presents itself as a conflict between the resident population and tourists, but also as a degradation of the cultural and environmental heritage of a destination, as well as in the form of distortion of the labour market (Milano et al., 2022). Overtourism, in other words, has to do with all those events congesting the territory, risking compromising the relationship between tourists and the resident population, damaging the cultural and environmental assets that a destination has and creating low-quality and poorly paid employment by taking away labour from more productive resources.

Although the phenomenon is difficult to outline precisely, it is possible to identify some areas to observe in order to identify the dynamics of overtourism.

The first and most relevant, embracing in some way the multiple perspective employable, is the well-being and quality of life of the local communities involved. The interaction between local communities and tourists can often be conflictual, because the economic advantages are accompanied by processes of congestion, loss of local identity and the birth of conflicts for the use of scarce resources. Excess tourism gives rise, especially in cities of art, to phenomena of gentrification, or the transformation of particularly attractive destinations from places of stable residence of the inhabitants, into places of excessive concentration of accommodation facilities because they are more profitable. The increase in the prices of housing and other related services and goods empties the historic centres also leading to phenomena of loss of cultural identity.

Overtourism can also be declined in terms of seasonality of demand. The presence of tourists is not uniform during the different phases of the year, subjecting the territories to excessive pressure during some periods and leaving the accommodation facilities unused in others. This phenomenon, linked to the organization of production, the habits of the population and the need to conduct some activities (schools, universities, etc.) during some time intervals, also greatly influences the quality of employment which is composed of a large number of jobs with fixed-term contracts and few stable jobs.

Visitors and tourists have an impact on transport. A high intensity of tourist flows requires an increase in networks to the areas visited, an increase in internal connections and a subtraction of mobility spaces from the resident population.

Overtourism leads to excessive exploitation of environmental resources and excessive waste production. The issue of protecting the environment by minimizing the use of its resources and recycling the waste produced is an issue that concerns the entire resident population. However, when a large number of tourists are added to the territory, it is difficult to manage the waste cycle and the recycling path in a sustainable way, perhaps for a small municipality or for an area already congested by a large population.

Overtourism can also be expressed in terms of conservation of cultural and environmental sites. Conserving and giving value to cultural and natural sites represents the most important challenge to ensure the harmonious growth of tourism. They are, in fact, the prerequisite for the arrival of tourists, but they must be preserved and enhanced to ensure their usability also for future generations.

Overtourism therefore involves economic, social and environmental aspects and can be identified as the threshold above which tourism is not anymore beneficial for economic growth in a given territory (Jordan et al., 2018).

This threshold is not univocally identified as it depends on the carrying capacity and the specific features of each territory (Milano, 2018; Weber et al., 2017). It is therefore a phenomenon that can be better evaluated from a microeconomic perspective, despite some indicators might have more general validity. It needs to be defined in space and time in order to be able to outline the characteristics of the destination and follow its path through time. Below some indicators capable of capturing at least some general aspects of the phenomenon of overtourism are presented.

5.2 Tourism Pressure Indicators: Alternative Perspectives

The different perspectives from which to observe the phenomenon of overtourism have led scholars to develop a variety of indicators, each one aimed at capturing a different aspect. In order to define each indicator, it is important to circumscribe the space and time to which they refer.

They can capture information from the tourist demand side (presences, arrivals, etc.) or from the supply side (number and occupancy of accommodation facilities) or capture the pressure on the territory with respect to specific aspects (waste or pollution generated by tourism, number of visitors to an archaeological or cultural site). Indicators are always relative measures, in the sense that absolute numbers must always be related to a reference phenomenon. Here are some of the indicators, aware that none of them can capture overtourism as a whole and that many furthers could be elaborated.

The first one is:

1. Tourism density:

$$\text{TD} = \frac{P_t}{\text{KM}^2} \quad (5.1)$$

where tourist density (TD) is calculated as the ratio between tourist presences at time $t - P_t$—and the square kilometres—KM^2—of the spatial surface that hosted them. It is therefore an indicator of territorial congestion that refers to the physical space of a city, region or country, etc. The higher this ratio is, the greater the risk of overtourism. If calculated for each month of the year, it will be able to provide information regarding the seasonality of tourist demand.

2. Tourism intensity

$$\text{TI} = \frac{P_t}{\text{POP}_t} \quad (5.2)$$

Tourism intensity (TI) is expressed as the ratio between the presences and the population in a given territory and in a given timespan. This indicator therefore relates the phenomenon of tourism to the quality of life of the resident population. Even the tourist intensity indicator, if calculated month by month, provides information on the seasonality phenomenon.

Compared to the tourist density indicator, which captures information that has to do with the physical size of the territory, tourist intensity catches the possible presence of conflicts with residents in that territory and the impact of the presence of too many tourists on the quality of life of citizens. The two indicators observe the same phenomenon from two different perspectives, one physical and the other social. However, it might happen that as the presences in the territory increase, the physical space is equipped to manage the many tourists or that the resident population receives benefits more than damages from the interactions with visitors. The result depends on what described in the previous chapter, that is, the preferences of the community, the ability of policymakers to manage tourism and any economic and non-economic benefits generated by tourism activities, such as increased income, increased employment or simply the creation of a more welcoming and livelier climate in the territory. To account for this complexity, there is an indicator that captures the contemporary interaction of tourist presences with both the territory and society (Canale & De Siano, 2021; De Siano & Canale, 2022, 2024):

3. Tourism territorial pressure

$$\text{TTP} = \frac{P_t}{\text{POP\_DENS}_t} \qquad (5.3)$$

where P_t are as usual the presences and time t and POP_DENS_t represents the number of inhabitants per square kilometre both in a given space (Canale & De Siano, 2021; De Siano & Canale, 2022, 2024).

This indicator is aimed at capturing the interaction of tourism with the territory of a given province taking into account how tourism flows affect the pre-existing congestion level and their contribution to the per capita income growth of residents. The advantage of this indicator is its capacity to detect the natural vocation of the territories, distinguishing between those relying on environmental "gifts", and therefore characterized by a lower population density, and those endowed with more cultural and archaeological resources endowed of intense past and present population settlements. This distinction defines also the attitude of residents to welcoming tourists and deriving income from directly and indirectly connected activities. Therefore, it can occur that—given the number of presences—an area with a low population density registers a higher territorial pressure with respect to one with a higher number of residents per square km (and vice versa). This is the result of a calculation strategy that, despite being technically simple, captures the complex phenomenon of interaction between tourism flows, the territory and its contribution to per capita income. This indicator complies with the criteria presented in the previous chapter combining economic results and local preferences to assess sustainability. In fact, this indicator returns values that are not uniquely defined, but rather depending on the specific features of each territory, residents' attitude to "bear" tourism presences as well as economic results.

Further specific indicators capturing overtourism are referred to specific phenomena. Here are some selected ones.

4. Gross utilization index

$$\text{GUI} = \frac{P_t}{\text{BP} * \text{dd}} \quad (5.4)$$

where P_t is the usual number of presences in a period of time, BP is the number of beds available in the accommodation facilities in a territory and dd is the number of days during which calculate the index.

This indicator captures overcrowding by observing the phenomenon from the supply side perspective. It shows the percentage of actual use of available beds, providing a measure of the intensity with which accommodation facilities are used. High use may signal excessive pressure on facilities, contributing to overcrowding. The greater the index the higher the potential overtourism: it ranges from zero to one meaning that when it is at the lower bound the accommodations have been always empty during the considered period (dd) while when it reaches the value of 1 all the accommodations are full for all the days. It is an indicator of potential overtourism as fully occupied accommodation structures do not necessarily mean that tourists are too much for the territory.

5. Air transport density

$$\text{AirD} = \frac{\text{Airports}}{\text{KM}^2} \quad (5.5)$$

where AirD is the number of airports per squared km. It is intuitive that the higher the number of airports in a given territory, the higher the potential influx of tourists. It should be noted, however, that airports usually serve a rather large territory and therefore this indicator is not always able to capture information on a phenomenon that should instead be investigated at the level of specific destinations. However, as Peeters et al. (2018) suggest an indicator that better captures overtourism in relation to airports is

6. Air travel density:

$$\text{ATD} = \frac{\text{Pass}_t}{P_t}$$

where Pass_t stands for "passengers" and P_t indicates as usual the presences in a given territory and time. The higher the ratio the higher the degree of overtourism. This indicator is able to capture the number of passengers flying for a very short stay, therefore overcrowding territories and producing CO_2 emissions while giving back very little to the local population.

7. City tourism congestion index

$$\text{CTCI} = \frac{A_{st}}{A_{tot}} \qquad (5.6)$$

CTCI is the ratio between the number of short-stay accommodation establishments A_{st} and the total accommodations establishments A_{tot} of the city under examination. This indicator—particularly suitable for cities of art—considers how much of the tourist demand is satisfied by facilities that offer short-term stays such as B&Bs compared to the total number of facilities. It therefore captures hit-and-run tourism and indirectly also how much the homes for residents are transformed into accommodation facilities. It is an indicator of overtourism, because it signals the expulsion of residents from the city centre due to the effect it generates on the availability of accommodations and their price.

8. Share of tourism in the economy

$$\text{TS} = \frac{\text{GDP}_{\text{TOUR}}}{\text{GDP}_{\text{TOT}}} \qquad (5.7)$$

This indicator that expresses the contribution to tourism GDP as a percentage of the total GDP of a territory is of particular importance and requires reflection. Without a doubt, in fact, a high contribution to tourism GDP signals its importance in the economic structure of a country. However, this circumstance is also accompanied by numerous presences in the territory and possible negative impacts on the population resulting from the excessive use of services or the production of pollution or waste that local institutions are unable to manage. The tourism sector is also characterized by fragility with respect to the occurrence of exogenous shocks (wars, terrorism, pandemics, etc.) or sudden decline in the attractiveness of a destination that could cause a collapse of the economic activity of the territory. These features are also connected with the phenomenon of resilience, or the recovery capacity of a region linked to the rapid recovery of the tourism sector that will be addressed in more detail in the following pages. Finally, the tourism sector is characterized by low added value, exposing firms and workers of the sector to low profitability and wages. The indices presented above capture only a part of the potential information relating to overtourism and an exhaustive treatment is not possible, given that researchers and institutions, depending on the phenomenon of their interest, continually develop new ones: therefore, the reader is left with a general framework of tools from which to start for any further investigations.

It is clear, however, that the phenomenon of overtourism can be observed from a multitude of perspectives concerning both the nature of the destination and the impact on the society. The following pages will try to account for the different perspective offered by some of the indicators reported above and to show how the results do not always converge towards a single conclusion.

5.3 The Phenomenon of Overtourism and the Different Nature of the Territories

The phenomenon of overtourism takes different forms depending on the indicator used and the territory to which it refers. To account for this complexity, some of the indicators presented above are calculated, with the aim of comparing one with another and observing from a practical point of view how they vary through time. The first set of indicators is the most general one referring to the territory as a whole and are TD, TI and TTP.

Figure 5.1 presents the dynamics of tourism density—the ratio between the number of presences and the squared kilometres on which they occurred—in selected European countries from 2005 to 2019, the year before the pandemics (the years during the pandemics have been excluded for the nature and extraordinary exogenous shock). The countries considered are those covering the most important role in the tourism sector in Europe, namely France, Germany, Greece, Italy and Spain.

The value of the indicator in the denominator is constant (KM^2), therefore the figure also shows the increase in presences over time. They grow for almost all countries starting from 2010 (years following the negative shock of the financial crisis) with a more marked increase for Italy in recent years. The country less affected by overtourism—according to this indicator—is Greece, while Italy is the most affected one. Italy is a small country in terms of size of the territory compared to the others, but it welcomes tourist flows from all over the world thanks to the length

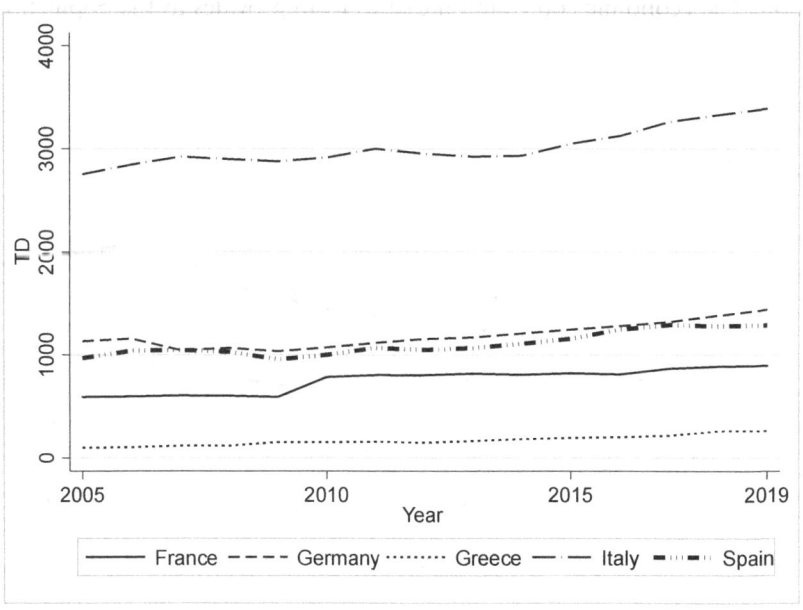

Fig. 5.1 Tourism density in selected European Countries (*Source* Own elaboration on Eurostat data)

of the coasts to the cultural and historical heritage and to the variety of destinations (seaside and mountains destinations and cities of art).

When examining the behaviour of the indicator "tourism intensity" over the same countries (Fig. 5.2) the evaluation of the phenomenon of overtourism changes. Greece jumps to first place with a considerable increase starting from the financial crisis and even more starting from 2013 (certainly due to the price competitiveness and migrations both consequence of austerity policies imposed to restore public budgets). Germany slips to last place while Italy moves to third. Common to all countries, although to a different extent, is the growth of tourist arrivals.

Finally, when examining the indicator "tourism territorial pressure" it is possible to perceive how, when considering the social and economic interaction of tourism with the territory and the population, things change again (Fig. 5.3). It is true that Greece remains the most affected by overtourism, but the surprise is Italy, which, having a high population density, is characterized by low congestion for the entire period under consideration. The economic contribution that tourism provides to the population

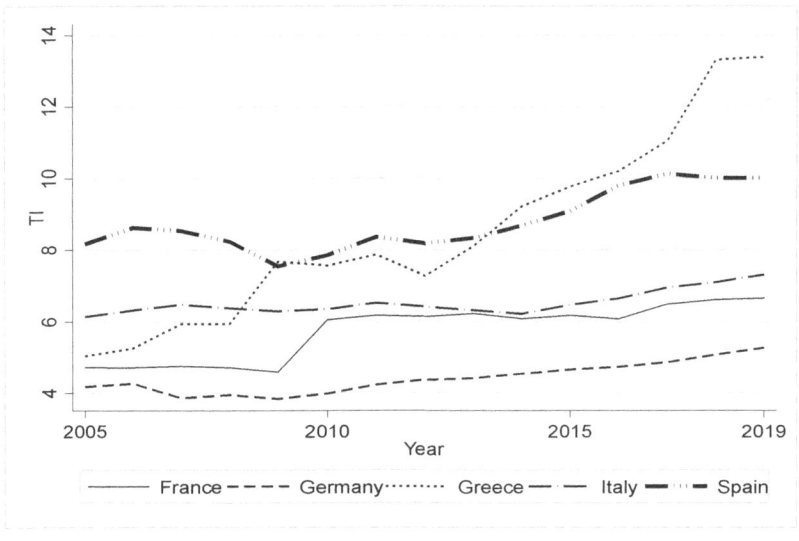

Fig. 5.2 Tourism intensity in selected European Countries (*Source* Own elaboration on Eurostat data)

further composes the puzzle, tipping the balance in the direction of greater benefits or greater costs in each of the countries considered.

The comparison between these indicators provides a first understanding of the complexity of the phenomenon of overtourism and how even for very large territories, such as entire countries, the dynamics can be very different depending on the observation perspective. When proceeding in depth by analysing the characteristics of individual destinations that cover smaller territories and have peculiar characteristics, it is possible to grasp phenomena that are not captured at an aggregate level.

This is the case of art cities affected especially in recent years by the proliferation of short-stay accommodation establishments. As stated in relation to overtourism in cities, the growth of short-term accommodations increases costs for residents, progressively expelling them from cities and distorting the characteristics of historic centres. The phenomenon is so relevant that the statistical office of the European Union (Eurostat) started to collect data in European cities. This information allows to

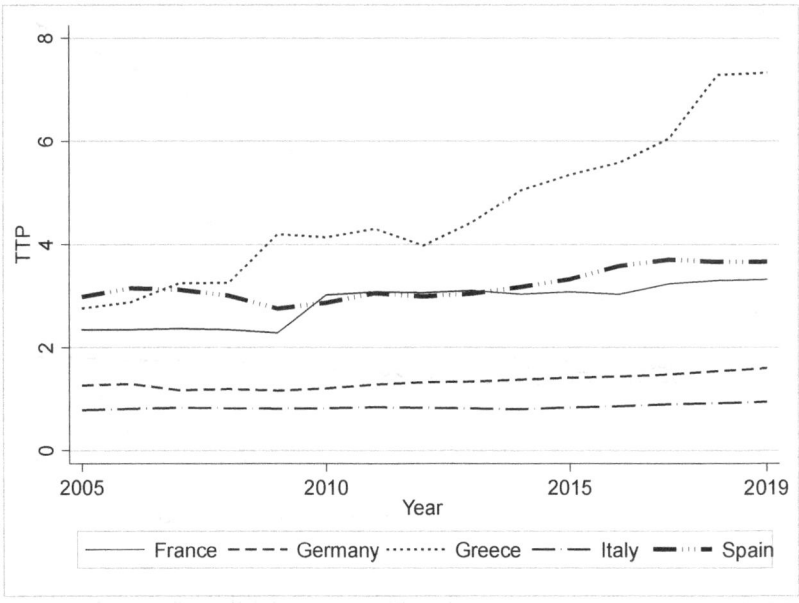

Fig. 5.3 Tourism territorial pressure (ml) in selected European Countries (*Source* Own elaboration on Eurostat data)

provide a broad picture of the phenomenon from 2020 to 2023 in six cities of art (Fig. 5.4).

From almost the end of pandemics it is a growing phenomenon in the five cities considered—Athens, Berlin, Madrid, Paris, Praha and Rome. However, Athens Rome and Madrid are the most affected by the increase in presences in structures such as B&B.

When having a look to the dynamics of the indicator ST—short-stay accommodation establishments as percentage of total accommodation establishments (Fig. 5.5)—the city most affected by overtourism is Athens, while the last affected is Rome. Furthermore, it is observable that the values of the indicator are rather stable in all cities considered, revealing that the fast increase in short stays in the last two years depicted in Fig. 5.4 has been obtained at expenses of total accommodations. The changing nature of tourism provides information about the potential consequences on the territory, both in regard to the impact on

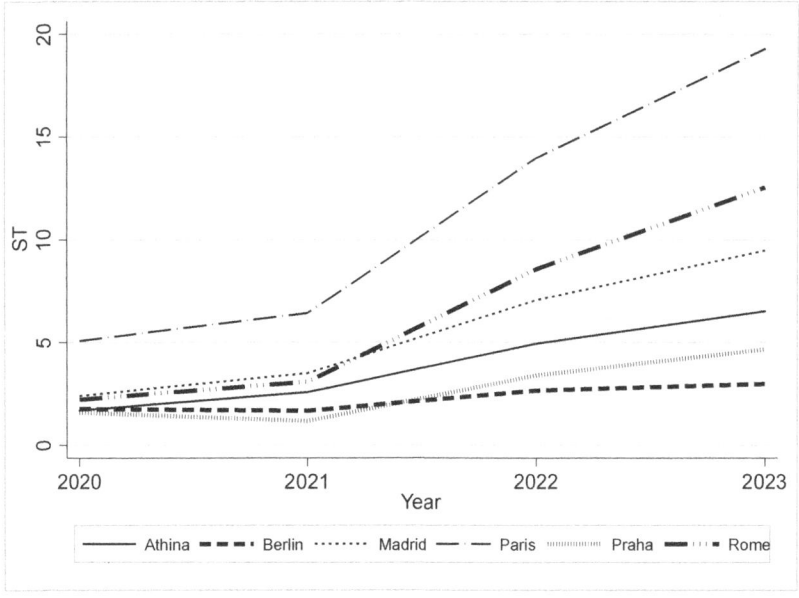

Fig. 5.4 Short-stay accommodation offered via collaborative economy platforms in selected touristic European cities (*Source* Own elaboration on Eurostat data)

the population and the size of investments in the sectors, that is, about the qualitative evolution of the tourism sector.

From the examination of the trend over time of some of the indicators proposed in the previous paragraph, the evolution of the phenomenon of overtourism can be assessed from different perspectives and that there is no univocal measure of the phenomenon. Furthermore, there is no threshold beyond which it can be established unambiguously that a territory is affected by overtourism. Evaluations by researchers and policymakers should be carried out by cross-referencing data at an aggregate level, with data obtained at the highest possible level of detail, investigating at the regional (NUTS2), provincial (NUTS3) and individual city levels, also accounting for preferences and needs of the communities hosting tourists (Peeters et al., 2018). Finally, distinctions in evaluating overtourism should be made between advanced, developing and underdeveloped economies.

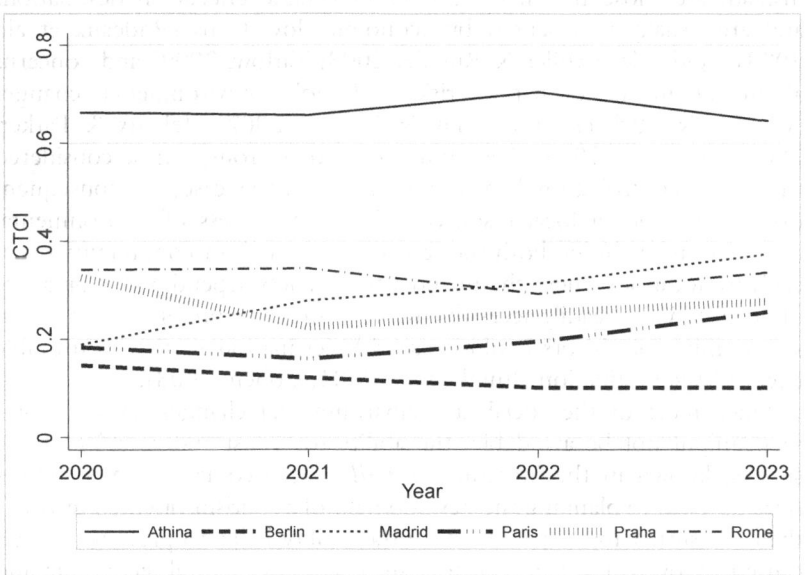

Fig. 5.5 City tourism congestion index in selected touristic European capitals (*Source* Own elaboration on Eurostat data)

5.4 Resilience and Tourism

The close interdependence between tourism and the territory, which takes the form of both direct and indirect reciprocal influences with all the activities present in it, means that tourism and all the players involved (residents, businesses and visitors) are particularly exposed to changes and shocks of various kinds (climatic, demographic, economic, etc.).

The vulnerability of tourist destinations to socio-economic shocks and stressors, such as environmental calamities or those related to the phenomenon of mass tourism as demonstrated in the previous paragraphs, has been extensively studied in the economic literature (Sharpley, 2005). The factors that contribute most to aggravating the vulnerability of a tourist destination are those with rapid onset that directly affect the region, such as political unrest, terrorist attacks, environmental and natural disasters causing a sudden drop in its attractiveness (Biggs et al., 2012; Mansfeld, 1999; Richter & Waugh, 1986). Slower onset factors, instead, are those that may have unpredictable effects on destinations and are usually represented by economic downturns (Prideaux et al., 2003), epidemics (Miller & Ritchie, 2003; Tarlow, 2009) and concerns about the impact of natural risks and global environmental changes (Cioccio & Michael, 2007; Hay & Becken, 2007; Méheux & Parker, 2006; Orchiston, 2013). Last, but no less dangerous, can be considered the out-of-control growth of tourist flows. In this case, the consequent overexploitation of local resources triggers a process of environmental degradation damaging both the territory and the host community. Especially for locations where charm and attractiveness depend on the presence of protected and readily accessible heritage sites, the presence of an excessive number of visitors could cause the destination to the irretrievable decay of the destination (Burak et al., 2004; Cohen, 2008).

Since most of the social and environmental changes, even if foreseen, might not be avoidable, the ability to "resist and react" to these shocks, known in the literature as *resilience*, becomes the more effective approach to planning the development of a tourism-oriented territory than the sustainability paradigm. Sustainability helps to prevent changes caused by overexploitation of tourism resources (natural, economic and cultural), ensuring that uses do not exceed the regenerative capacity of the resources. Resilience, instead, helps to develop the capacity to adapt to changes, both expected and unexpected, and increases the ability of a destination to renew itself when the shocks undermine the sustainability of its main activities.

Resilience may depend on several factors. First, the interconnectedness between social, economic and ecological components; second, the diversity that allows for a reduction in the incidence of risks; and, finally, the flexibility required to respond promptly to shocks and foster the development of the destination (Folke et al., 2003; Nelson et al., 2007).

In a study published in 2014, Lew reconstructed a classification which associated different levels of resilience and tourism context depending on whether the shock was sudden or not and whether it involved mainly private entrepreneurs or shared public interests (Lew, 2014). This classification is presented in Fig. 5.6 where four possible scenarios may arise on based on the degree of disturbance (x-axis—from gradual shift to sudden shock) and the scale of tourism actors (y-axis—from private entrepreneurs to shared public interests).

Lew's model assumes that individual entrepreneurs handle changes in the environment, culture and society differently depending on whether these are slow or severe and unexpected. The response to the shock

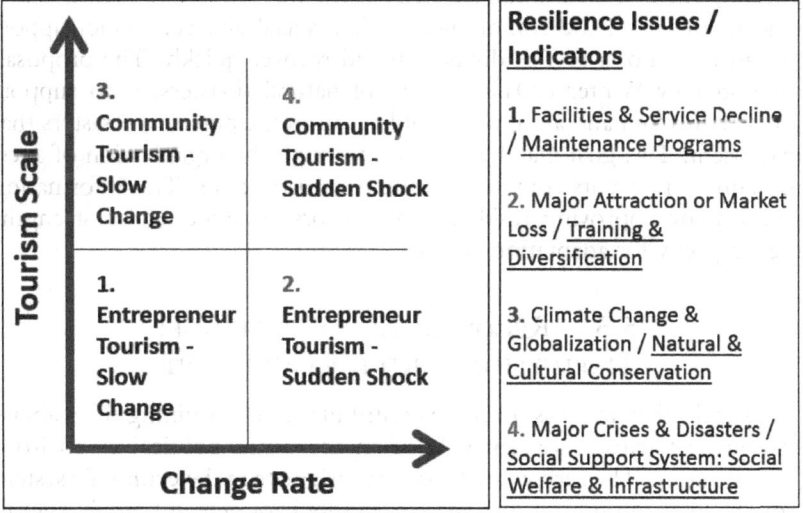

Fig. 5.6 Scale, change and resilience (SCR) in tourism (*Source* Lew [2014])

also varies depending on the dynamics in time, space or social structure involved.

The first quadrant is the one related to gradual changes affecting individual tourism entrepreneur. In this case, entrepreneurs try to modify declining structures and services to meet the changing needs of tourists to safeguard only their own profitability. If the gradual change affects the entire community of the tourist destination (third quadrant), the latter tries to provide for the conservation of the natural and cultural heritage through, for example, green certifications for hotels or other types of tourist services, and corporate social responsibility practices, i.e. the set of behaviours and practices adopted by entrepreneurs to achieve common interest goals haw could water conservation, fossil fuel consumption and cultural resource preservation be considered. In presence of sudden event, such as a flood or an economic crisis, the main concern of each entrepreneur is the loss of access to the main tourist attraction resources or to the main tourist markets due to infrastructural, economic, political or other disruptions. In this context, the most effective strategy might be achieving an adequate level of diversification of both customers and suppliers (second quadrant). The last scenario concerns sudden shocks affecting the entire community, such as economic crises and natural disasters. In this case, the community needs a social and economic support system to respond to this disruption and recover quickly. The proposal, suggested by Winter (2011) in case of natural disasters, is to support public education and awareness-building on all the potential disasters that may occur. This goal may be achieved through the organization of sites, museums and events commemorating past disruptions. This information pathway, by improving residents and visitors' awareness, also strengths their capacity for adaptation and recovery.

5.5 Recovery and Resilience: Definitions and Measurement

The shocks that are most likely to contribute to undermining the stability of a given territory are generally of an economic, political or environmental nature. The former are those that originate in the economic system from sudden and unexpected changes in supply or demand, or changes in financial markets, technology and government policies. Although initially limited to a specific sector of the economy, due to the close interdependence of markets and industries, economic shocks propagate causing broad macroeconomic changes. Political shocks, instead, refer to changes in the international or domestic systems affecting processes, relationships and expectations influencing interactions between states, nations or individual actors within a country. Political shocks can be either exogenous, such as conflicts, or endogenous, such as changes in administrative or political regimes that usually follow a sharpening of internal social contrasts. Finally, environmental shocks are basically those that can be attributed to natural disasters such as floods, droughts, typhoons, earthquakes, volcanic eruptions, tsunamis or to environmentally damaging events caused by human activity such as pollution, deforestation and nuclear accidents, among others.

Changes following whichever shocks can be negative, which means that they can be detrimental to the territory, or positive, thus rising new opportunities for development and growth. However, the greatest concern, for policymakers and scholars, is for negative shocks because they undermine the collective capacity of individuals, organizations and communities to tolerate, absorb or adjust to changes that affected them (Nelson et al., 2007). This capacity depends mainly on three factors. The first is adaptation, understood as the decision-making process and the set of actions put in place to cope with current or future expected changes. The second is the degree of vulnerability, i.e. the sensitivity of a system to any perturbation, which depends, in turn, on the adaptive capacity. The last is what is referred to in the literature as *"resilience"* (Holling, 1973; Martin, 2012; Nelson et al., 2007) that is the capacity of a system to respond positively to changes, maintaining its main functions despite the impact of a negative shock.

The concept of resilience was first introduced by Holling (1973) with reference to ecology, as the ability of ecosystems to survive disturbances or adverse changes in environmental conditions. Later, it has been extended to other fields and enriched by a pool of indicators and measurement methods.

In the economic context, resilience is defined as the ability of a local economy to respond to and recover from recessionary shocks or unexpected adverse changes affecting aggregate supply and demand, technological level or government policies (Crescenzi et al., 2016; Lagravinese, 2015). Three main concepts of economic resilience may be found in the literature (Martin, 2012; Martin & Sunley, 2015). The first, the *engineering resilience* graphically represented in Fig. 5.7, refers to the resistance of a system, that is the ability and the speed at which it returns to its pre-shock equilibrium (Holling, 1973; Pimm, 1984; Walker et al., 2006). This ability implies the presence of a self-correcting mechanism that helps to keep production and socio-economic structures unchanged.

However, since market frictions could prevent the system from reacting promptly, causing a lack of resilience, two scenarios may arise as shown in Fig. 5.8:

(a) a permanent drop in level, while recovering the pre-shock growth rate;
(b) a permanent drop in both level and rate of growth.

If able to react, instead, regions may eliminate unproductive activities and open new strategic sectors with better development opportunities bringing them to higher levels and/or trends growth paths. (Bonß, 2016; Muštra et al., 2016). Shifts presented in Fig. 5.9 show the presence of multiple equilibria, depending on system's own elasticity to changes (*hysteresis*, as defined by Martin, 2012):

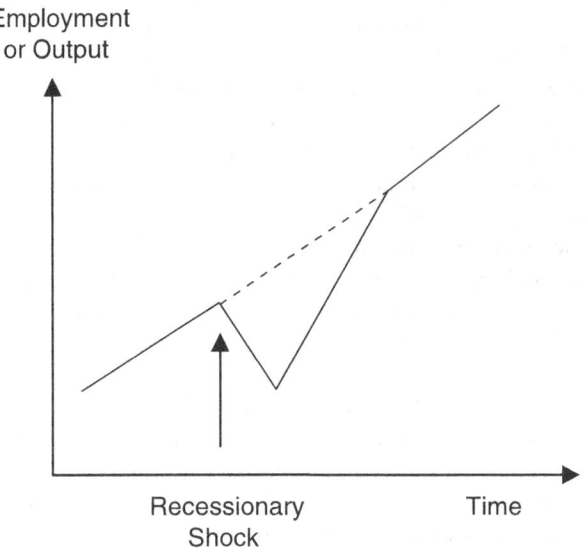

Fig. 5.7 Impact of a recessionary shock: region returns to pre-shock trend (*Source* Martin and Gardiner [2019])

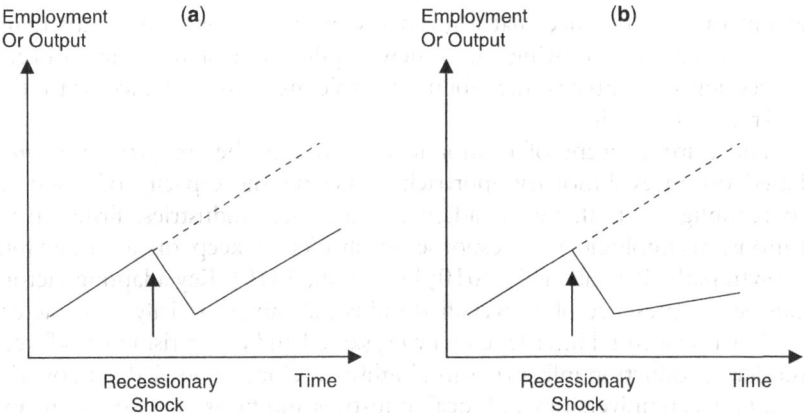

Fig. 5.8 Impact of a recessionary shock: lack of resilience (*Source* Martin and Gardiner [2019])

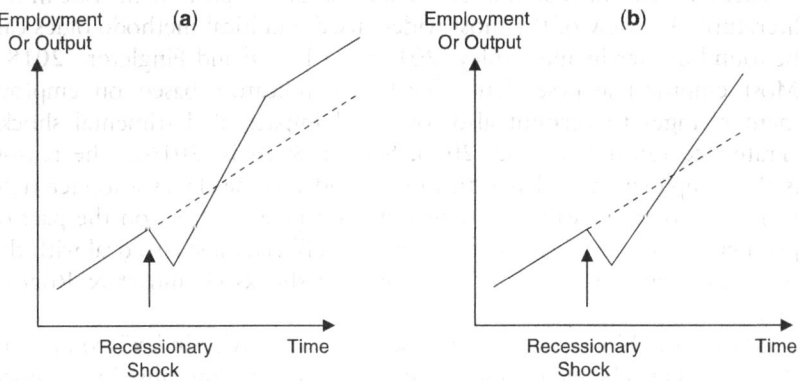

Fig. 5.9 Impact of a recessionary shock: possible positive outcomes (*Source* Martin and Gardiner [2019])

(a) a permanent rise in level with pre-shock growth rate.
(b) increasing levels and growth rate compared to the pre-shock period.

The latter scenario leads to the second definition of economic resilience provided by the literature, the *ecological resilience*, which defines the extent of a disturbance that a given system can absorb before changing its structure and moving to a new equilibrium state. This concept of resilience combines the ability to tolerate a disturbance with the self-renewal capacity.

The third concept of economic resilience is the *adaptive resilience*, based on an evolutionary approach, denoting the capacity of a system to reconfigure itself, i.e. to adapt its structure (industries, firms, institutions, technologies) in response to shocks to keep on a sustainable growth path (Pendall et al., 2010; Pike et al., 2010). Key adapting factors can be the presence of a diversified industrial mix, especially if characterized by low sectoral interdependencies, skilled and enterprising workforce, modern production infrastructure, highly developed knowledge networks (i.e. between universities and local industries and firms), supportive financial systems, liberal market conditions and political dynamism (Boschma, 2015; Christopherson et al., 2010; Davies & Tonts, 2010; Desrochers & Leppälä, 2011; Di Caro, 2014; Martin, 2012; Sensier & Artis, 2016).

Measurement of economic resilience is a further prominent issue in the literature. A review of the most widely used empirical methodologies can be found in Martin and Sunley (2015) and Doran and Fingleton (2018). Most empirical analyses have developed indicators based on employment changes to account also for social impact of detrimental shocks (Fratesi & Rodriguez-Pose, 2016; Sensier & Artis, 2016). The reason is that employment, relative to output and income, takes a longer time to return to pre-crisis levels. This implies a greater effort on the part of planners who, in addition to the economic effects, must also deal with the social problems raised by severe economic shocks (Reinhart & Rogoff, 2009).

Martin (2012) suggested the use of a "sensitivity index" comparing the percentage change in employment in a given region caused by a shock with the employment change caused by the same shock at national level. Denoting by E the level of employment in each geographical area (region, country), the sensitivity index β_r (where subscript r denotes the region) would then look as follows:

$$\beta_r = \frac{\Delta E_r / E_r}{\Delta E_N / E_N} \tag{5.9}$$

The sensitivity index is therefore greater than one if the region is less resilient to a negative shock compared to the average country level, while it is less than one if the region is more resilient (low sensitive to shocks) compared to the national average.

Regarding tourism, the literature presents numerous theoretical and empirical contributions on resilience of tourist destinations after stressful events. These ex-post analyses consist of comparing individual indicators of tourism sector performance before and after the shock. The most frequently used indicators include the number of people employed in the tourism industry, the number of accommodation facilities in each destination, the number of tourists arrivals, the number of overnight stays and the level of aggregate product, to name a few. The generic resilience index would then be:

$$\text{Resilience Index (RI)} = \frac{\text{Indictor after the shock}}{\text{Indicator under normal coditions before the shock}} \quad (5.10)$$

A resilience index (RI) greater than 1 indicates that the geographical unit is resilient, conversely, an index less than 1 means lack of resilience. The higher the RI, the greater the resilience.

Table 5.1 and Fig. 5.10 present values of the resilience index of the tourism sector, built using the number of "total stays at the short-stay accommodations offered via collaborative economy platforms", to the Covid-19 pandemic shock. The index uses annual data from Eurostat statistics[1] for the six European capitals, considered in the previous paragraphs, over the period 2018–2023.

Index values presented in Table 5.1 and the graph in Fig. 5.10 show that in all the considered cities the pandemic by Covid-19 represented a devasting shock also from the heath but also from the tourism flows perspective. Despite the size of the collapse, all cities showed a quick recovery of the tourism sector with a more than doubled value of the index just one year after the removal of restrictions. Rome more than the other cities strongly recovers after 2020, Athens in Greece seems to have a rapider recover at the beginning but just as quickly slowed the course of the recovery; a slower reaction has been observed for Berlin and Praha.

Usually, previous empirical investigations on tourism resilience have been focused on the recovery by tourism industries as well as arrivals

---

[1] https://ec.europa.eu/eurostat/databrowser/view/tour_ce_oarc/default/table?lang=en.

Table 5.1 Resilience index using total stays at short-term accommodation in European cities

City	2019	2020	2021	2022	2023
Praha	1.08	0.25	0.74	2.88	1.37
Berlin	1.02	0.42	0.94	1.59	1.12
Athina	1.29	0.37	1.52	1.91	1.32
Madrid	1.13	0.30	1.46	2.01	1.34
Paris	1.02	0.38	1.26	2.18	1.38
Roma	1.11	0.22	1.39	2.78	1.46

Source Own elaboration on Eurostat dataset

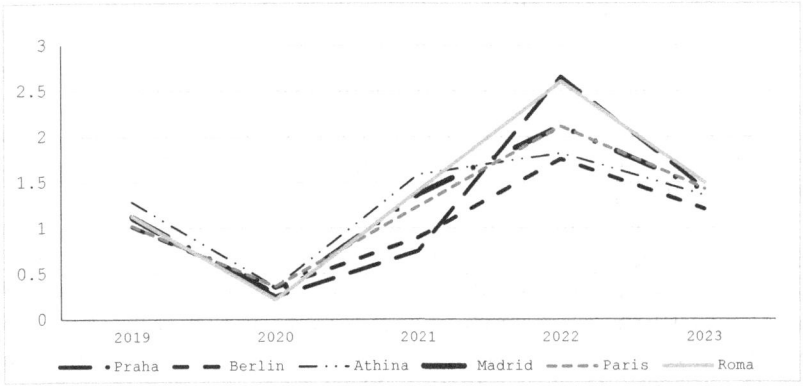

Fig. 5.10 Resilience index for total stays after the pandemic by Covid-19 (*Source* Own elaboration on Eurostat dataset)

from rapid disruptions caused by disasters and crises (Faulkner, 2000), such as the Asian economic crisis of the late 1990s (Pearce, 2001), the SARS epidemic in China in 2002–2003 (Zeng et al., 2005), the Indian Ocean tsunami in 2004 (Biggs et al., 2012; Calgaro & Lloyd, 2008; Smith & Henderson, 2008), or combinations of economic, political and business crises (Lew, 1999; Prideaux et al., 2003, for Southeast Asia). For a comprehensive overview see Ritchie (2004, 2009) and Hall et al. (2013). More recently, slow-changing variables have begun to receive some attention from tourism scholars who started relating resilience to Butler's (1980) TALC model (as in Hamzah & Hampton, 2013;

Petrosillo et al., 2006) or evaluating impacts of economic migration and social change on tourist destinations (Cheer and Lew, 2018). Finally, another strand of literature has considered the resilience of destinations with respect to global climate change (Becken & Hay, 2007; Kajan & Saarinen, 2013). The aims of these studies were, for example, the effects on tourism behaviour (Becken & Wilson, 2013) or on specific tourism industries, such as those related to winter sports in mountain resorts (Steiger & Stötter, 2013) and scuba diving services in reef areas (Biggs, 2011; Hillmer-Pegram, 2013), driven by the ongoing climate change.

5.6 Reconciling Contrasting Impacts: The Critical Resilience Threshold of a Tourist Destination

Equally important is the process of analysis that leads to defining the resilience threshold of a tourist destination, interpreted as the maximum level of tourism-driven resource exploitation or stress that a site/territory can sustain, beyond which the characteristics of the tourist activity are profoundly modified to the point of making it impossible to practice. In an urban tourism context, for example, the depletion of these resources could completely undermine the attractiveness of the location, thus wiping out not only visitors willing to pay more but also those included in different categories. Similarly, in the case of ecotourism, the exploitation of key resources, such as rare wildlife species and pristine environments, may entail the risk of their extinction. In these and other cases, the main problem is tourism planners' incapacity to identify the correct resilience threshold at an early stage, which means allowing available resources to be exploited even beyond sustainable limits.

The literature on early assessment of critical thresholds for resilience in the tourism sector is rather limited. In what follows, the approach suggested by Johnston and Tyrrel (2008) is described. The authors, starting from a model of sustainable tourism (Johnston and Tyrrel, 2005), refer to the dynamic of tourism resources consumption to identify the scritical resilience threshold of a given destination.

This approach suggests identifying the amount of tourism resources that can be used to maximize the net total benefit (NTB) of tourism activities in each location, the latter equal to the sum of the net benefits of residents and tourism enterprises. Given the optimum size of resource exploitation, and if each visitor consumes a fixed quantity of resources in each period, the optimum number of visitors ensuring tourism sustainability can be easily calculated. By considering tourism resources as renewable, tourism sustainability implies a trade-off between the number of visitors, and hence the number of resources they use, and the regenerative capacity of resources for each initial endowment. Visitors, in fact, even if attracted by high levels of resources, use part of them during their stay, thus contributing to a deterioration of the quality of the environment. Resource endowments, however, may change due to their regenerative capacity and tourists' exploitation. Labelling with X the amount or tourism resources, $h(X)$ its dynamic and V the amount consumed by each tourist, equal to 1 for simplification. Environmental quality dynamics may be as follows:

$$\dot{X} = h(X) - V \qquad (5.11)$$

where $\dot{X}$ indicates the change in the stock of available tourism resources. The condition of sustainability in presence of tourism flows is reached when

$$\dot{X} = 0 \qquad (5.12)$$

hence

$$h(X) = V \qquad (5.13)$$

According to Johnston and Tyrrel the quality of a destination depends on several factors. More precisely they assume the existence of three dimensions of quality, each determined by ecological-environmental (Xn), economic-fiscal (Xe) and socio-cultural (Xc) resources. Government strategies, size and timing of their interventions, together with destination management organizations choices may affect the environmental quality and attractiveness of the destination, which must be considered when searching for the optimal size of resources exploitation. Finally, destination size, resources endowments and social infrastructure may also influence the choice of the sustainable number of tourists a destination may afford.

The usefulness of this model is to provide analytical tools such as the variation of natural resource growth rates, visitor perceptions of change and thresholds for the risk of collapse that have to be considered when choosing the optimal policy to achieved industry profits, high quality of life and employment in the tourism sector. This information is used by planners to decide how and when to control the number and type of tourists accessing a destination both directly, through targeted advertising and promotions, and indirectly through investments in environmental resources. The objective is to continue along a path of sustainability of tourism activities over time.

REFERENCES

Albaladejo, I. P., & Gonzalez-Martínez, M. (2019). Congestion affecting the dynamic of tourism demand: Evidence from the most popular destinations in Spain. *Current Issues in Tourism, 22*(13), 1638–1652. https://doi.org/10.1080/13683500.2017.1420043

Balaguer, J., & Cantavella-Jordà, M. (2002). Tourism as a long-run economic growth factor: The Spanish case. *Applied Economics, 34*(7), 877–884.

Becken, S., & Hay, J. E. (2007). *Tourism and climate change: Risks and opportunities*. Channel View.

Becken, S., & Wilson, J. (2013). The impacts of weather on tourist travel. *Tourism Geographies, 15*(4), 620–639.

Biggs, D. (2011). Understanding resilience in a vulnerable industry: The case of reef tourism in Australia. *Ecology and Society, 16*(1), 30.

Biggs, D., Hall, C. M., & Stoeckl, N. (2012). The resilience of formal and informal enterprises to disasters: Reef tourism in Phuket, Thailand. *Journal of Sustainable Tourism, 20*, 645–665.

Bonß, W. (2016). *The notion of resilience: Trajectories and social science perspectives*. In A. Mauer.

Boschma, R. (2015). Towards an evolutionary perspective on regional resilience. *Regional Studies, 49*(5), 733–751.

Burak, S., Dogan, E., & Gazioglu, C. (2004). Impact of urbanization and tourism on coastal environment. *Ocean & Coastal Management, 47*(9–10), 515–527.

Butler, R. W. (1980). The concept of a tourist area cycle of evolution: Implications for management of resources. *Canadian Geography, 24*(1), 5–12. https://doi.org/10.1111/j.1541-0064.1980.tb00970.x

Calgaro, E., & Lloyd, K. (2008). Sun, sea, sand and tsunami: Examining disaster vulnerability in the tourism community of Khao Lak, Thailand. *Singapore Journal of Tropical Geography, 29*, 288–306.

Canale, R. R., & De Siano, R. (2021). Territorial pressure and tourism contribution to GDP: The case of Italian regions. *International Journal of Tourism Research, 23*(5), 891–900. https://doi.org/10.1002/jtr.2451

Cheer, J. M., & Lew, A. A. (2018). Understanding tourism resilience. *Tourism resilience and adaptation to environmental change: Definitions and frameworks.* New York: Routledge.

Christopherson, S., Michie, J., & Tyler, P. (2010). Regional resilience: Theoretical and empirical perspectives. *Cambridge Journal of Regions, Economy and Society, 3*(1), 3–10.

Cioccio, L., & Michael, E. J. (2007). Hazard or disaster: Tourism management for the inevitable in Northeast Victoria. *Tourism Management, 28*(1), 1–11.

Cohen, E. (2008). The tsunami waves and the paradisic cycle: The changing image of the Andaman coastal region of Thailand. *Tourism Analysis, 13,* 221–232.

Crescenzi, R., Luca, D., & Milio, S. (2016). The geography of the economic crisis in Europe: National macroeconomic conditions, regional structural factors and short-term economic performance. *Cambridge Journal of Regions, Economy and Society, 9*(1), 13–32.

Davies, A., & Tonts, M. (2010). Economic diversity and regional socioeconomic performance: An empirical analysis of the Western Australian grain belt. *Geographical Research, 48*(3), 223–234.

De Siano, R., & Canale, R. R. (2022). Controversial effects of tourism on economic growth: A spatial analysis on Italian provincial data. *Land Use Policy, 117.* https://doi.org/10.1016/j.landusepol.2022.106081

De Siano, R., & Canale, R. R. (2024). The role of tourism in European regions' economic recovery: A spatial perspective. *Tourism Economics, 30*(8), 2021–2042. https://doi.org/10.1177/13548166241248679

Desrochers, P., & Leppälä, S. (2011). Creative cities and regions: The case for local economic diversity. *Creativity and Innovation Management, 20*(1), 59–69.

Di Caro, P. (2014). Recessions, recoveries and regional resilience: Evidence on Italy. *Cambridge Journal of Regions, Economy and Society, 8*(2), 273–291.

Doran, J., & Fingleton, B. (2018). US metropolitan area resilience: Insights from dynamic spatial panel estimation. *Environment and Planning A: Economy and Space, 50*(1), 111–132.

Dritsakis, N. (2012). Tourism development and economic growth in seven Mediterranean countries: A panel data approach. *Tourism Economics, 18*(4), 801–816.

Durbarry, R. (2004). Tourism and economic growth: The case of Mauritius. *Tourism Economics, 10*(4), 389–401.

Ehigiamusoe, K. U. (2020). Tourism, growth and environment: Analysis of non-linear and moderating effects. *Journal of Sustainable Tourism, 28*(8), 1174–1192. https://doi.org/10.1080/09669582.2020.1729164

Fahimi, A., Saint Akadiri, S., Seraj, M., & Akadiri, A. C. (2018). Testing the role of tourism and human capital development in economic growth. A panel causality study of micro states. *Tourism Management Perspectives, 28*, 62–70.
Faulkner, B. (2000, August). *"The future ain't what it used to be"*: Coping with change, turbulence and disasters in tourism research and destination management. Professorial Lecture Series No. 6. Paper presented at Griffith University, Queensland, Australia.
Folke, C., Colding, J., & Berkes, F. (2003). Building resilience and adaptive capacity in social-ecological systems. In *Navigating social-ecological systems* (pp. 352–387). Cambridge University Press.
Fratesi, U., & Rodríguez-Pose, A. (2016). The crisis and regional employment in Europe: What role for sheltered economies? *Cambridge Journal of Regions, Economy and Society, 9*(1), 33–57.
Hall, C. M., Timothy, D. J., & Duval, D. T. (Eds.). (2013). *Safety and security in tourism: Relationships, management, and marketing*. Routledge.
Hamzah, A., & Hampton, M. P. (2013). Resilience and non-linear change in island tourism. *Tourism Geographies, 15*(1), 43–67.
Hay, J., & Becken, S. (2007). *Tourism and climate change: Risks and opportunities*. Channel View Books.
Hillmer-Pegram, K. C. (2013). Understanding the resilience of dive tourism to complex change. *Tourism Geographies, 16*(4), 598–614.
Holling, C. S. (1973). Resilience and stability of ecological system. *Annual Review of Ecology and Systematics, 4*, 1–23.
Johnston, R. J., & Tyrrell, T. J. (2005). A dynamic model of sustainable tourism. *Journal of Travel Research, 44*(2), 124–134.
Johnston, R. J., & Tyrrell, T. J. (2008). Sustainability and tourism dynamics. In *Tourism management: Analysis, behaviour and strategy* (pp. 470–492). CAB International.
Jordan, P., Pastras, P., & Psarros, M. (2018). *Managing tourism growth in Europe*. The ECM toolbox. Dijon keep.Eu. Search by countries & regions. https://www.keep.eu/keep/nuts/searchByRegion#null
Kaján, E., & Saarinen, J. (2013). Tourism, climate change and adaptation: A review. *Current Issues in Tourism, 16*(2), 167–195.
Lagravinese, R. (2015). Regional resilience and rising gaps North-South: The case of Italy. *Cambridge Journal of Regions, Economy and Society Advance, 8*, 331–342.
Lew, A. A. (1999). Tourism and the Southeast Asian crises of 1997 and 1998: A view from Singapore. *Current Issues in Tourism, 2*(4), 304–315.
Lew, A. A. (2014). Scale, change and resilience in community tourism planning. *Tourism Geographies, 16*(1), 14–22.

Mansfeld, Y. (1999). Cycles of war, terror, and peace: Determinants and management of crisis and recovery of the Israeli tourism industry. *Journal of Travel Research, 38*(1), 30–36.

Martin, R. (2012). Regional economic resilience, hysteresis and recessionary shocks. *Journal of Economic Geography, 12*(1), 1–32.

Martin, R., & Sunley, P. (2015). On the notion of regional economic resilience: Conceptualisation and explanation. *Journal of Economic Geography, 14*, 1–42.

Martin, R., & Gardiner, B. (2019). The resilience of cities to economic shocks: A tale of four recessions (and the challenge of Brexit). *Papers in Regional Science, 98*(4), 1801–1833.

Méheux, K., & Parker, E. (2006). Tourist sector perceptions of natural hazards in Vanuatu and the implications for a small island developing state. *Tourism Management, 27*(1), 69–85.

Milano, C. (2018). Overtourism, malestar social y turismofobia. Un debate controvertido. PASOS. *Revista de Turismo y Patrimonio Cultural, 6*(3), 551–564.

Milano, C., Novelli, M., & Cheer, J. M. (2022). Overtourism. In D. Buhalis (Ed.), *Encyclopedia of tourism management and marketing*. Edward Elgar Publishing. https://doi.org/10.4337/9781800377486.over.tourism

Miller, G. A., & Ritchie, B. W. (2003). A farming crisis or a tourism disaster? An analysis of the foot and mouth disease in the UK. *Current Issues in Tourism, 6*(2), 150–171.

Muštra, V., Šimundić, B., & Kuliš, Z. (2016). *Effects of smart specialization on regional economic resilience in EU*. First SMARTER Conference on Smart Specialisation and Territorial Development: Changing Patterns of Territorial Policy: Smart Specialisation & Innovation in Europe.

Nelson, D. R., Adger, W. N., & Brown, K. (2007). Adaptation to environmental change: Contributions of a resilience framework. *Annual Review of Environmental Resources, 32*, 395–419.

Orchiston, C. (2013). Tourism business preparedness, resilience and disaster planning in a region of high seismic risk: The case of the Southern Alps. New Zealand. *Current Issues in Tourism, 16*(5), 477–494.

Pearce, D. (2001). Tourism. Asian. *Pacific Viewpoint, 42*(1), 75–84.

Peeters, P., Gössling, S., Klijs, J., Milano, C., Novelli, M., Dijkmans, C., Eijgelaar, E., Hartman, S., Heslinga, J., Isaac, R., Mitas, O., Moretti, S., Nawijn, J., Papp, B., & Postma, A. (2018). *Research for TRAN Committee—Overtourism: Impact and possible policy responses*. European Parliament, Policy Department for Structural and Cohesion Policies, Brussels.

Pendall, R., Foster, K. A., & Cowell, M. (2010). Resilience and regions: Building understanding of the metaphor. *Cambridge Journal of Regions, Economy and Society, 3*(1), 71–84.

Petrosillo, I., Zurlini, G., Grato, E., & Zaccarelli, N. (2006). Indicating fragility of socio-ecological tourism-based systems. *Ecological Indicators, 6*(1), 104–113.
Pike, A., Dawley, S., & Tomaney, J. (2010). Resilience, adaptation and adaptability. *Cambridge Journal of Regions, Economy and Society, 3*, 59–70.
Pimm, S. L. (1984). The complexity and stability of economic system. *Nature, 307*, 321–326.
Poa, W., & Huang, B. (2008). Tourism development and economic growth—A nonlinear approach. *Physica A: Statistical Mechanics and Its Applications, 387*(22), 5535–5542. https://doi.org/10.1016/j.physa.2008.05.037387
Prideaux, B., Laws, E., & Faulkner, B. (2003). Events in Indonesia: Exploring the limits to formal tourism trends forecasting methods in complex crisis situations. *Tourism Management, 24*, 475–487.
Reinhart, C. M., & Rogoff, K. S. (2009). The aftermath of financial crises. *American Economic Review, 99*(2), 466–72.
Richter, L. K., & Waugh, W. L. (1986). Terrorism and tourism as logical companions. *Tourism Management, 7*(4), 230–238.
Ritchie, B. W. (2004). Chaos, crises and disasters: A strategic approach to crisis management in the tourism industry. *Tourism Management, 25*, 669–683.
Ritchie, B. W. (2009). *Crisis and disaster management for tourism.* Channel View.
Santamaria, D., & Filis, G. (2019). Tourism demand and economic growth in Spain: New insights based on the yield curve. *Tourism Management, 75*, 447–459.
Sensier, M., & Artis, M. (2016). The resilience of employment in Wales: Through recession and into recovery. *Regional Studies, 50*(4), 586–599.
Sharpley, R. (2005). The tsunami and tourism: A comment. *Current Issues in Tourism, 8*(4), 344–349.
Smith, R. A., & Henderson, J. C. (2008). Integrated beach resorts, informal tourism commerce and the 2004 tsunami: Laguna Phuket in Thailand. *International Journal of Tourism Research, 10*(3), 271–282.
Steiger, R., & Stötter, J. (2013). Climate change impact assessment of Ski Tourism in Tyrol. *Tourism Geographies, 15*(4), 577–600.
Tarlow, P. E. (2009). *How the next pandemic may impact the world's tourism industry.* Tourism Review On-line Magazine IX/2009.
Walker, B., Gunderson, L., Kinzig, A., Folke, C., Carpenter, S., & Schultz, L. (2006). A handful of heuristic and some propositions for understanding resilience in socio-ecological systems. *Ecological and Society, 11*(1), 13.
Weber, F., Stettler, J., Priskin, J., Rosenberg-Taufer, B., Ponnapureddy, S., Fux, S., Camp, M.-A., & Barth, M. (2017). *Tourism destinations under pressure. Challenges and innovative solutions.* Lucerne University of Applied Sciences and Arts, Institute of Tourism ITW.
Winter, C. (2011). First world war cemeteries: Insights from visitor books. *Tourism Geographies, 13*(3), 462–479.

Zeng, B., Carter, R., & De Lacy, T. (2005). Short-term perturbations and tourism effects: The case of SARS in China. *Current Issues in Tourism, 8*(4), 306–322.

CHAPTER 6

Tourism, Innovation and Sustainability in Europe

Abstract The chapter proposes a practical perspective for examining the relationship between tourism and sustainability in advanced economies with a particular focus to Europe. European institutions propose a measurement system that, by involving the public and private sectors, investigates the various aspects of sustainability influenced by the tourism activities (ETIS). The diffusion of this methodology, applied for some time by numerous national and international organizations, is proposed as a first step towards the establishment of rules and policy indications consolidated at European level to make sustainable tourism a real tool for territorial development. The role of innovation in the tourism sector is investigated to evaluate its contribution to sustainable development and to economic growth in destination territories.

Keywords Tourism sustainability · Indicators of sustainability · Certifications · ETIS · Tourism innovation

6.1 The Dynamics of the Tourism Sector in the Main European Countries and the Issue of Sustainability

Destinations that have experienced, or are experiencing, steady growth in tourist arrivals are increasingly facing problems of crowding, localized inflation and pressure on accommodation, thus confirming what has been theorized by studies on the life cycle of tourist locations. The consequence is mistrust, and often even aversion, to a tourism-led growth process. Understanding the social impacts of tourism on communities is therefore necessary for governments to prevent a sentiment of hostility of local population against tourists and tourism development itself. More sustainable tourism pathways, in terms of environmental, social and economic effects, require two main conditions: the involvement of all stakeholders in tourism decision-processes, especially resident communities, and availability of basic information on the determinants and ways in which tourism activities expand in the territory, together with tools to measure their effects.

In this regard, this chapter addresses the issue of sustainable tourism from a more practical point of view. Having defined the theoretical framework and the complexity of the choices necessary to guarantee a balance between the environment and society, both in the present and in the future, it is now necessary to move on to explore how to implement these principles so that tourism can be the driver for sustainable growth. There is a clear interest of businesses and institutions to achieve this goal which ensures the consolidation of activities, the maintenance of employment levels and profitability of tourism over time. Besides, as awareness of environmental and social issues grows, sustainability becomes a market strategy to target tourists with more complex and conscious interests in their destinations.

The tourism sector in Europe represents a suitable context for testing strategies developed by scholars and institutions, addressing risks and seizing the opportunities provided by the extraordinary growth of what has become the "industry" of tourism. In European countries, in fact, tourism has demonstrated not only resilience in the face of economic and geopolitical challenges (De Siano & Canale, 2024), but also solid performance, as shown by the most recent data confirming that pre-pandemic flows have been exceeded. The latest research from the European Travel Commission (ETC) revealed that on the 3rd quarter of 2024 the growth of the sector led to a 7% rise in foreign arrivals and a 5% in overnight stays year-on-year.

Travel plays an essential role in human lives, despite the growing incidence of transport and services costs, and Europe is the first destination with Germany France, Italy and Spain among the ten most visited countries in the world. Observing the trend in tourist arrivals at tourist accommodation[1] in a sample of European countries (Fig. 6.1) a clear difference between Germany, France, Spain and Italy and the rest of the sample can be seen. All countries show an increasing trend from 2007 (the year of the global financial crisis) that collapses due to the pandemic crisis in 2020. However, except for Germany and UK, all countries show a restart of tourism as early as 2021 and in 2023 arrivals reach pre-pandemic flows.

Difference in flows and effects of the pandemic are even more evident in Fig. 6.2. While Germany strengthens its lead over time, the United Kingdom, that in 2015 attracted more tourists than France and Italy, is overtaken by these two countries in 2023. A contributing factor may have been the more difficult access to British territory after Brexit.[2] The flows characterizing the other countries in the sample appear negligible, although they are the most visited after the top 5. Even for them, the flows in 2023 significantly exceed those of the pre-pandemic period.

The dynamics of the tourism sector appears different when considering the supply side by measuring the number of establishments (Fig. 6.3). Among the sample countries, Italy's reception capacity is far greater than that of any other in Europe. One-third of the establishments, in fact, are located in Italy with the United Kingdom, Spain, Germany and France left at considerable distance. The structure of European accommodation establishments is dominated by holiday and other short-stay accommodation (NACE group 55.2), accounting for 69.9% of all establishments. Hotels and similar accommodation (NACE group 55.1) account for 26.2% and camping grounds (NACE group 55.3) for 3.9%. More than 40% of the total number of bed places available in Europe are in establishments located in coastal areas.

---

[1] Hotels; holiday and other short-stay accommodation; camping grounds, recreational vehicle parks and trailer parks.

[2] Brexit is the name given to the United Kingdom's departure from the European Union that took place on 1st of February 2020 after the referendum held on the 23rd of June 2016.

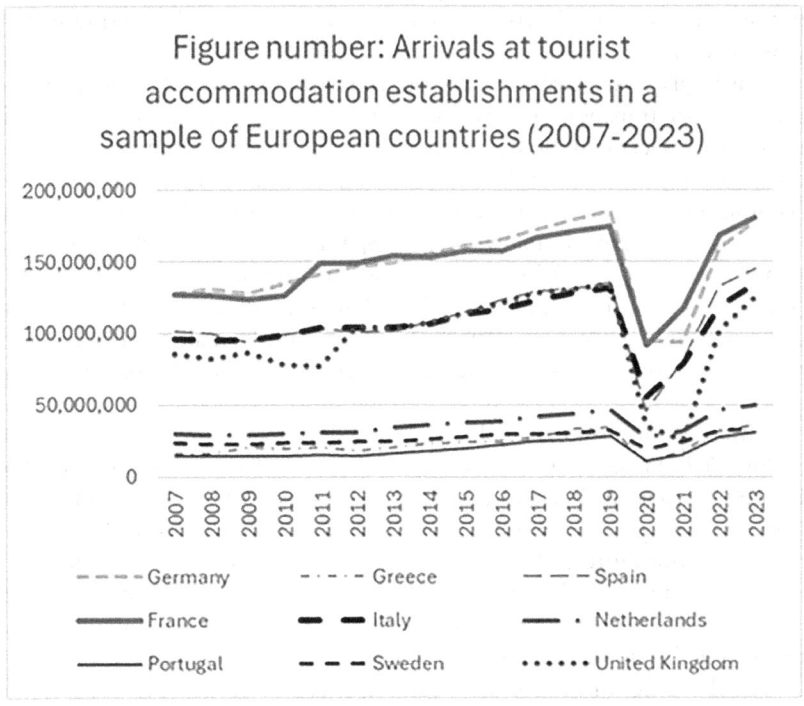

Fig. 6.1 Arrivals at tourist accommodation establishments in a sample of European countries (2007–2023) (*Source* Own elaboration on Eurostat datasets)

Many of Europe's destinations are facing problems of overtourism and need strategies and new tourism policies to address capacity constraints in most popular hotspots, relieve pressure on overcrowded areas by a redistribution of visitors and achieve a more equitable reallocation of benefits. To investigate the phenomenon of tourism sustainability is not enough looking at the number of arrivals at accommodations, but the suggestion is to consider rather its ratio to the size of the territory or population. To this end the number of arrivals at accommodation per 1000 inhabitants is calculated and presented in Fig. 6.4 and described in what follows.

If tourism is observed from this point of view, the ranking is almost completely reversed, placing Greece in first place, with a much stronger pressure of tourism on resident population, followed by Sweden, Spain,

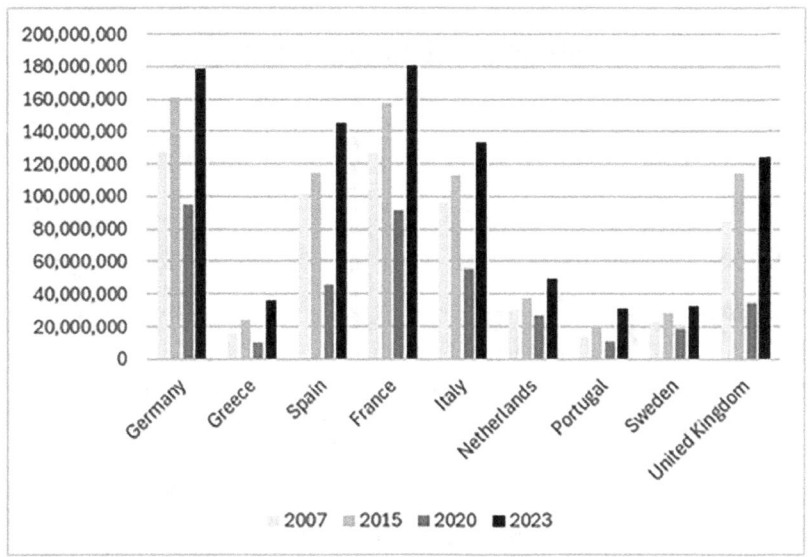

Fig. 6.2 Arrivals in selected years (*Source* Own elaboration on Eurostat datasets)

Portugal and the Netherlands. This indicator captures both the relationship with GDP per inhabitant and the degree of congestion. A high number of arrivals relative to the number of inhabitants, in fact, on the one hand indicates a greater economic benefit for the population but, on the other hand, also reveals greater congestion for the country.

There are different ways to evaluate sustainability, and it would be necessary to look at the phenomenon from all points of view as comparing arrivals, arrivals per inhabitant and arrivals per square kilometre, or tourism territorial pressure, may produce different results in terms of the impact flows may exert in each country.

6.2 Tourism and Sustainability in the European Union: The Institutional Framework

Within the European Union the responsibility for the tourism sector falls to national, regional and local authorities. However, with the adoption into law of the Lisbon Treaty (December 2009), the European Union

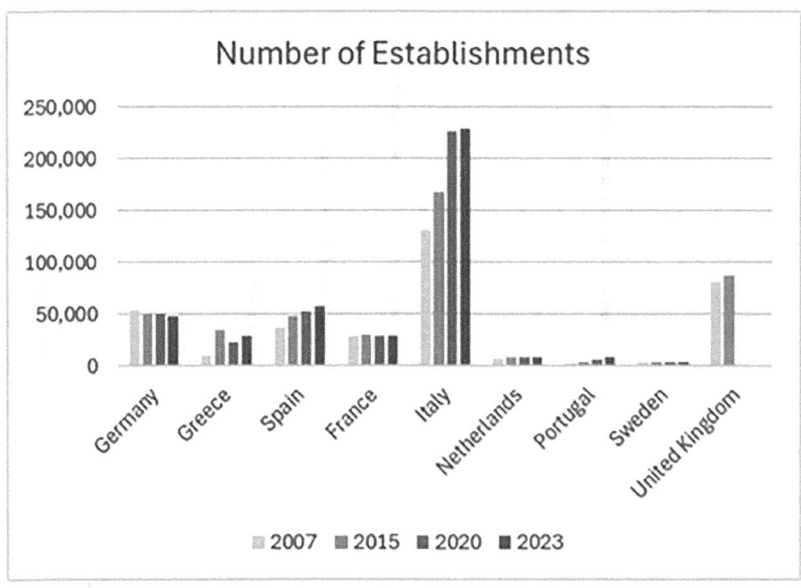

Fig. 6.3 Number of establishments in selected years (*Source* Own elaboration on Eurostat datasets)

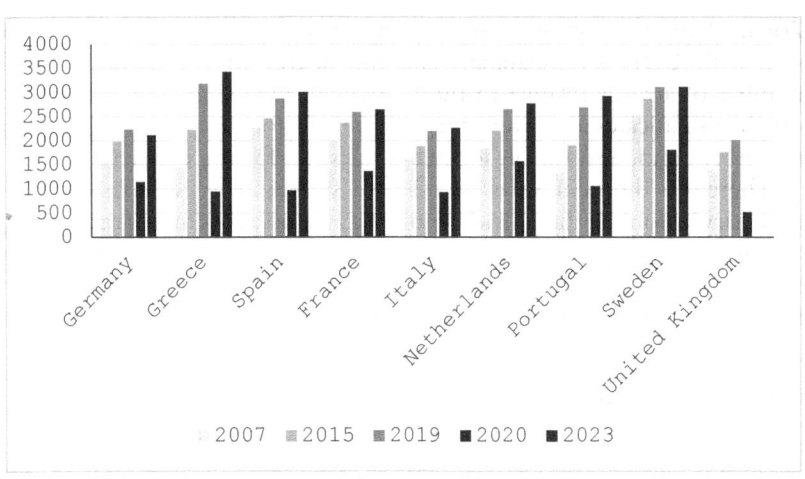

Fig. 6.4 Arrivals at tourism accommodations per 1000 inhabitants in European countries (*Source* Own elaboration on Eurostat datasets)

acquired specific power to direct tourism activities to respond to challenges and grasp opportunities facing the sector. Indeed, even if excluding any form of harmonization at European level regarding tourism, the treaty provides the Parliament and the Council with new legal powers to develop a coherent tourism policy approach aimed at strengthening the competitiveness of the sector. To this end, the Treaty stipulates that no new decision on tourism policy can be taken without the assent of the European Parliament. In detail, Article 195 of the Treaty states that "...*The European Parliament and the Council, acting in accordance with the ordinary legislative procedure, shall establish specific measures to complement actions with the Member States to:*

> *a) encouraging the creation of a favourable environment for the development of undertakings in this sector;*
> *(b) promoting cooperation between Member States, particularly by the exchange of good practice".*

These good practices were part of a broad strategy defined in the Europe2020 strategy, "Smart, inclusive and sustainable growth" adopted by the European Council in 2010 (European Commission, 2010). The strategy defined three priorities for growth in European countries: *Smart growth* (developing an economy based on knowledge and innovation), *Inclusive growth* (fostering high employment and ensuring social and territorial cohesion) and *Sustainable growth* (promoting a more resources efficient, greener and more competitive economy).

Because of its structure, tourism is the economic sector in which the guiding principles of a smart, sustainable and inclusive Europe can be realized most quickly. It comprises businesses in several sectors, including food and beverage services, online information and services providers (e.g. tourist offices or digital platforms), travel agents and tour operators, accommodation suppliers, destination managing organizations, attractions and passenger transport (such as airlines and airports, trains, busses and boats). These are precisely activities that in recent decades have experienced extraordinary growth and a transition towards sustainability supported by numerous process and product innovations.

European countries are characterized by notable disparities, even though, because of the globalization and the interconnectedness of its value chains, tourism represents an important part of the EU's overall economy. The tourism policy approach at EU level, therefore, aims to maintain Europe's position as a leading global destination while completing its transition towards sustainability ensuring that benefits become as widespread as possible. To this end, the tourism development sought should be characterized by four main aspects:

- Stimulate competitiveness in the European tourism sector (consolidation of knowledge, uptake of the use of high-tech information systems, development of professional skills and improvement of the quality of the EU tourism offer).
- Promote the development of sustainable, responsible and high-quality tourism.
- Consolidate the image and profile of Europe by improving the tourist offer and the quality of services.
- Maximize outcomes of policies implemented by the EU with reference to the financing instruments made available by the centralized budget.

These goals, which seem to be declined in a too general way, take the form of a series of challenges to be met that balance tourism sector growth with the environmental, social and economic aspects of its sustainability.

The first is the seasonality of demand. The presence of tourists is not uniform during the year, therefore putting excessive pressure on the territories during some periods and leaving accommodation facilities unused during others. This phenomenon, linked to the organization of production, the habits of the population and the need to conduct certain activities (schools, universities, hospitals, etc.) during certain time intervals, also greatly affects the quality of employment, mainly composed of people with fixed-term contracts and few stable jobs. Reducing the seasonality of demand, and therefore spreading visitors throughout the year, means making tourism more sustainable by reducing pressure on the territory (environmental sustainability) and giving continuity to the production activities and receipt of incomes (economic sustainability).

The second is the impact of transport connected to tourist flows on the environment and territories. The intensity of tourist flows requires an increase in connections both towards and within destinations. The spread of low-cost flights, which has grown exponentially in recent decades, has made the aircraft the most widely used vehicle for travel, albeit the most polluting. The spread of low-cost flights, which has grown exponentially in recent decades, has made the aircraft the most widely used vehicle for travel, albeit the most polluting. The multiplication of internal transport has also contributed to the deterioration of environmental quality. The goal of tourism sustainability, hence, cannot be achieved without addressing the problem of environmental pollution caused by excessive transport.

The third is the quality of work in the tourism sector. Indeed, the tourism sector is characterized by part-time or temporary jobs and often still by undeclared work. Improving the quality of work means, by pushing in the direction of full-time jobs that are not dependent on the seasonality of demand, improving the economic contribution of tourism to growth and making it a stable component of GDP.

The fourth is the well-being and quality of life of the local communities involved. The interaction between local communities and tourists can often be conflictual, because economic benefits are increasingly accompanied by congestion, loss of local identity and disputes over the use of scarce resources. Sustainable tourism faces the challenge of maintaining a balance between the economic benefits and the social costs of overpopulation in the areas affected by the flows.

The fifth is minimizing the use of environmental resources and waste production by promoting recycling. This can become difficult to manage when many tourists are added to the territory, especially for a small municipality or an area already congested with a high population. The circular economy is not only a way to ensure the environmental sustainability of tourism resources, but also a way to produce raw materials in a cost-effective manner.

Preserving and valuing important cultural and natural sites is the most important challenge to ensure tourism sustainability. These sites are the main attractor of tourist flows and must be preserved to ensure their usability also for future generations and valorized for a stable contribution to the economic growth of the territory.

Finally, the last challenge is to make holidays accessible to all. Since 2005, the European Statistical Office (Eurostat) has included the economic possibility of taking a short holiday among the assets to calculate the level of poverty and social exclusion. The possibility of enjoying accessible holidays, by promoting integration and exchange between cultures and peoples, makes tourism also a tool to counter the competition generated by globalization that exposes people on the margins of society to the risk of isolation.

6.3 The Certification System and the Regional Network in Europe

The close connection between the competitiveness of tourism activity and its sustainability makes the sector an instrument for promoting the European vision of development. The European economic model, in fact, is largely based on the role played by small and medium-sized enterprises in the market, which thanks to their dynamism are able to respond to the continuous and changing needs of demand, ensuring employment also through the self-entrepreneurship, a necessary tool to cope with the progressive downsizing of the public sector and the growing need to ensure a high level of employment. In this regard, tourism has a great potential in terms of economic sustainability thanks also to its resilience that helps recover both for itself and the related sectors.

To help destinations measuring their performance in relation to sustainability, the European Commission in 2013 commissioned a group of experts—the Tourism Sustainability Group (TSG)—to create a European Tourism indicators system (ETIS, European Commission, 2016, 2019). The TSG, composed of public and private sector experts in the field of sustainable tourism, advises the European Commission on the design of policies to promote a more sustainable and competitive approach to tourism management and development. The indicator system allows a destination to be assigned a score for each of 4 specific areas to assess its performance in terms of tourism sustainability. Given its structure, the system supports destination in managing a sustainable development process, in collecting data and detailed information to monitor their performance from one year to another and lets policymakers and other stakeholders to have useful information on the level of sustainability of the destination.

The ETIS toolkit provides also guidelines and clear explanations on each indicator, both core and supplementary, and how to use them. The ETIS model (the complete set of indicators is presented in the appendix to this chapter) is composed of four pillars and 43 core indicators, together with an indicative set of supplementary indicators.

The four areas are summarized as follows:

(A) Destination management

A.1 Number of enterprises choosing sustainable tourism as a percentage of enterprises in the sector
A.1.1 Number of enterprises choosing to adopt environmental or social responsibility certification

(B) Economic value

B.1 Tourist flows arriving at the chosen destinationB.1.1 Number of nights at
B.2 Performance of tourism enterprises
B.3 Quantity and quality of occupancy
B.4 The tourism sector supply chain

(C) Social and cultural impact

C.1 Impact on the community
C.2 Health and safety
C.3 Gender equality
C.4 Inclusion and accessibility
C.5 Protection and enhancement of cultural heritage, local identity and assets

(D) Environmental impact.

D.1 Reducing the Impact of Transport
D.2 Climate Change
D.3 Solid waste management
D.4 Waste water management
D.5 Water Management
D.6 Energy use
D.7 Landscape and biodiversity management

Indicators are internationally comparable, which is an advantage because this allows tourism managers and policymakers to assess sustainability and improvements needed across the different elements of tourism using a common approach. Destination can choose for themselves which indicators to consider, which means greater flexibility and adaptability of the system to each context. Results are then based on self-assessment, observations, data collection and analysis by the destinations themselves. The critical issue emerged from experiences of implementing the ETIS system regards the collection of statistical data which would require technologies that perform better than those currently in use.

Many other initiatives support tourism sustainability taking the form of certifications for the activities and the supplied services.

The corporate social responsibility (CSR), among others, is defined by the European Commission (2011) as "*a guiding business policy whereby companies integrate social and environmental concerns in their own business mission, strategies and operation and in their interaction with their stakeholders on a voluntary basis*". It aims at promoting a corporate policy capable of harmonizing economic objectives with social and environmental objectives, with reference to the territory in which enterprises operate or with the aim of preserving the environmental, social and human heritage for the current and future generations. The idea is achieving business policy whereby tourism companies integrate social and environmental concerns in their own business mission, strategies and interaction with their stakeholders (employees, tourists, other businesses in the tourism supply chain, shareholders, investors, local communities, government authorities and media). Several basic international strategy papers are available to adopt CSR measures in tourism sector, some examples may be: the "*Global Code of Ethics for Tourism*", for the responsible and sustainable development of tourism; a Task Force established by the UNTWO to prevent sexual exploitation of minors, child labour and the trafficking of children and young people; the "*Davos Declaration*", suggesting mitigation strategies for tourism to respond to climate change.

With greater concern for environmental sustainability, the Eco-Management and Audit Scheme (EMAS) in the EU is a certification mechanism that takes place on a voluntary basis by tourism enterprises that want to achieve a better environmental performance and gain credibility and transparency in relation to the services offered (accommodation, food services, transport).

The European Union's ECO-label helps consumers to identify products and services that have a low environmental impact throughout their entire life cycle (from inputs' extraction, to use and disposal). The label is awarded by European institutions, with the collaboration of national institutions in the preliminary investigation phase which must indicate how the activity is managed, what type of energy is used, the amount of water used, management and recycling of waste together with criteria used to promote other good practices.

Furthermore, to promote sustainable tourism development models, EU uses a specific instrument labelled European Destination of Excellence (EDEN).[3] This is a competition, developed around an annual theme chosen by the European Commission in collaboration with national tourism boards, which aims to select and promote a tourism "destination of excellence" for each participating country. EDEN contributes to increasing the visibility of emerging and non-traditional European destinations, creating a platform for sharing good practices across Europe and promoting networking between awarded destinations.

Nowadays, the substantial question is whether the natural systems and changing climate conditions will be able to sustain tourism as an intense-resource-use economic sector. In this regard, the 7th EU Environmental Action Programme (7th EAP) gave the European Environment Agency (EEA) the task of monitoring worldwide initiatives aimed at measuring tourism environmental impacts and sustainability, as in the case of UNWTO's at a global level and INRouTe's[4] at subnational level. The latter has proposed a Regional Tourism Information System (R-TIS) to collect data at the national/regional levels for a limited number of items on specific aspects of tourism impacts: economic, territorial sustainable development and territorial cohesion. Most of these indicators are in line with the European Tourism Indicator System (ETIS). These include, among others, the attractiveness of places, water consumption, biodiversity disturbance, spread of sustainability practices through the adoption of environmental certification schemes and labelling, and potentials for ecotourism. Table A.6.1 in the appendix of the chapter (see Table A.6.1 in the Appendix) shows these indicators (Giulietti et al., 2018, p. 91).

Finally, World Tourism Organization (UNWTO) in collaboration with the Japan International Cooperation Agency (JICA) published in 2023[5] a toolkit of indicators applicable to tourism projects in international development settings aimed at strengthening the contribution of tourism to the achievement of the Sustainable Development Goals (SDGs). This

---

[3] https://ec.europa.eu/growth/sectors/tourism/eden/about_en.

[4] INRouTe is a private initiative promoted by UNWTO, aimed at measuring tourism and sustainable development at subnational levels, providing information for policy and analytical purposes, within the framework of the UN 2030 Agenda.

[5] Bonilla, J. (2023). *Achieving sustainable development goals through tourism: Toolkit of indicators for projects (TIPs)*.

toolkit, applied in tourism business, destination management, community development, academic research and public administration, helps in disseminating the value of tourism in creating better places for people to live in and visit. This toolkit provides indicators for the SDGs targets, for all the 17 SDGs, and indicates how these can be achieved through tourism projects whereas other existing measurement systems are mainly shaped to measure the impact of tourism on a specific area of sustainability in a determined level of governance or to assess the contribution of specific types of stakeholders.

6.4 Innovation in Tourism as a Mean to Reach the Objective of Sustainability: The Case of Europe

The competitiveness of the destination cannot be separated from the sustainability of tourism activities. The protection and preservation of natural resources, socio-cultural heritage and local cultures have become essential components of the tourism experience. Over time, there has been a growing awareness about the damages produced by the development models adopted so far, also in the tourism context, thus shifting the attention of policymakers and scholars to the impact and riskiness of the persistence of these production and consumption models.

The relationship between economic growth and environmental degradation, acknowledged by the literature as "Environmental Kuznets Curve" (EKC), suggests that innovation, and in particular eco-innovation, could represent a valuable tool to shape tourism further development and mitigate the environmental impact of its main activities such as transport and accommodation.

Innovative approaches should therefore be introduced in the tourism sector to ensure its growth and commercial success, on one side, and its economic, social and environmental sustainability, on the other side. The demand for innovative practices comes above all from tourism enterprises facing new technologies, new customers or changes in their consumption patterns, new competitors and, finally, new environmental needs and opportunities (green energy, sustainability, etc.). However, identifying innovations strictly related to tourism is difficult because of the size and dynamism of the sector which includes different activities, from accommodation, transportation, food and beverage services to retail trade. Innovations can meet a wide range of needs, from product creation to marketing campaigns for their promotion, with the aim of addressing consumer needs (Miralles, 2010) or strengthening the competitiveness, attractiveness and adaptation of enterprises to technological change (Hall & Williams, 2008; Razafindravelo, 2017). In this regard, digital transformation is playing a key role in keeping tourism enterprises' competitiveness and meeting the expectations of new and dynamic consumers. Internet, mobile and smartphone networks, big data and new cloud services, faster connectivity networks, social networks and artificial intelligence are just some tools contributing to this epochal change. Home automation and nanotechnology implemented in housekeeping practices, artificial intelligence and augmented reality in the museums are some of the opportunities innovative tools may provide to develop new tourism activities or revitalize and increase the attractiveness of the more traditional ones. Demand for digitalization arises along both the tourism value chain and the journey, from smart travel facilitation (smart visas, borders, security processes, infrastructures) to smart destination management (technology, sustainability, accessibility, inclusivity) and secure payment systems that make travelling more safety.

ICTs have thus provided new tools for the tourism industry, giving rise to new experiences for tourists. Likewise, the swift uptake of mobile technologies by tourists and visitors has enabled travellers to consume individualized information, regardless of the site and situation in which they are located. Leveraging technology makes it possible to personalize the customer experience and improve her/his satisfaction as issues can be dealt with in real time. Moreover, technology makes it possible to streamline operations and reducing costs in many departments.

Tourism managers, on their side, should work to make the digital transformation inclusive by promoting equitable access to digital tools and infrastructures and supporting training programmes for SMEs, particularly in rural areas and for vulnerable groups such as women, youth, indigenous peoples and persons with disabilities, and for the elderly. In this regard, the UNWTO, starting from the technology skills shortages highlighted during the pandemic crisis and the needs of tourism subsectors, developed a training programme for the digital transition of SMEs wishing to strengthen their competitiveness in their respective tourism destinations. The programme, called "Digital Futures",[6] includes diagnostic tools and digital pathways made available online for SMEs' easy engagement in further business development.

As tourism has all the characteristics of an integrated industry but also those of the service sector, such as the participation of the customer in the service process and the simultaneous production and consumption, tourism-related innovations focus on different areas such as product, process, service, technology, and social innovation as well as business and management.

There are three main innovations and technology trends in the travel and tourism industry aimed at improving visitors' experiences. The Near Frequency Communication technology (NFC) that is a form of wireless technology which allows two devices to essentially talk to each other and exchange data. It can be used to provide visitors with key information and to personalize their customer journey, as for example information about local attractions or good places to eat nearby without waiting to talk to the hotel concierge or creating self-guided tours for museums, art galleries, historical sites, parks, resorts and so on. Artificial intelligence (AI) is the way in which tourism businesses efficiently create personalized experiences for their guests. AI systems by collecting and processing huge data can customize visitors' experiences thus best fulfilling their expectations. Finally, virtual reality which is more often used by establishments such as hotels, cruises, vacation tours to improve the booking experience by including a preview of the experience they are about to book.

After the Covid-19 pandemic, the new EU industrial strategy emphasized the need to further accelerate green and digital transitions and increase the resilience of European industrial ecosystems. As the tourism was the most affected during the pandemic, it has been the first where a co-creation process was started by the Commission aimed at designing a transition pathway in line with the European Agenda for Tourism 2030/2050. The Commission launched a stakeholder consultation engaging 24 EU Member States, plus Norway and the United Kingdom, and in 2022 published a strategic roadmap for the tourism sector, identifying measures and outputs to accelerate the green and digital transitions and improve the resilience of the ecosystem and inviting tourism stakeholders to play their role in the process.

---

[6] "UNWTO Digital Futures Programme for Small and Medium-sized Enterprise (SMES)", online available at: https://www.unwto.org/digitalfutures.

The objectives of the European Green Deal need strategies favouring the efficient use of natural resources and the reduction of waste and pollutant emissions. In the tourism ecosystem these goals should be achieved while ensuring the health and safety of visitors. Key strategies and their related targets include:

- Passenger transport companies (aviation, water and land transport) should reduce greenhouse gases (GHG) in line with the European climate law goals to become climate neutral by 2050
- Cities and regions should develop sustainable mobility for climate-neutral cities, thanks also to the use of EU space data and applications, with sustainable mobility plans and by doubling the highspeed rail traffic by 2030, following the sustainable and smart mobility strategy and the EU mission "Climate Neutral and Smart Cities".
- Buildings have to become more energy and resource efficient, and energy support must move towards sustainable renewables, following the renovation wave strategy and its action plan.
- Hospitality and all tourism services should work towards circular models and reduce their environmental footprint, in line with the circular economy action plan and zero pollution action plan. This includes halving the amount of residual (non-recycled) waste by 2030 and reducing food waste, single-use plastics, air pollutants and water pollution while increasing water efficiency.
- Water efficiency needs to be improved by and by promoting the wider use of drought management plans as well as sustainable soil management and land use, in line with the climate adaptation strategy.
- Consumers' choices must be empowered, through more transparent information about the sustainability and environmental footprints of the destinations/tourism services, in line with the circular economy action plan.
- Coastal and maritime tourism actors should develop sustainable tourism in line with the new approach for the sustainable blue economy in the EU, European strategy for more growth and jobs in coastal and maritime tourism and the EU mission on "Restore our Ocean and Waters by 2030".

6.5 THE FUTURE OF TOURISM IN ADVANCED ECONOMIES: LIMITS, OPPORTUNITIES AND CHALLENGES

The tourism sector plays a very important role in the world economy as shown by the results achieved in the last decades, excluding the Covid-19 pandemic years, in most of the countries. The whole travel and tourism sector, with a share of 10.3% of global GDP and annual growth rates higher than those of the world economy, proves to be more and more a key sector for the economic growth. In the context of extraordinary growth of this sector, the last decades confirmed Europe as the main inbound region, with destinations in Mediterranean countries, like Italy, Greece, Portugal and Croatia, presenting even double-digit growth for tourism arrivals. Tourism contributions range from income and employment opportunities, increasing tax revenues, higher infrastructure investments, improved national balance of payments and environmental conservation and cultural heritage protection (Balaguer & Cantavella-Jordá, 2002; Dritsakis, 2012; Durbarry, 2004; Fahimi et al., 2018). This has been proven to occur also in structurally weak and peripheral areas where, beyond providing jobs and income, tourism contributed to attract capital, avoid outmigration and, on the contrary, favouring incoming migration more than other sectors. On the base of these contributions, tourism can be considered as a possible solution to uneven regional development and an opportunity for local communities' employment.

Alongside these contributions, however, there is also evidence that massive and uncoordinated influx of tourists to worldwide popular destinations may have strong negative environmental and social impacts. A degradation of the environmental context may be caused by and excessive exploitation of natural resources (water, air, soil), higher levels of waste production, energy use or human activities such as production, consumption, transport and urbanization. The social impact of tourism, instead, refers to the visitor-resident relations. A continued and rapid growth in tourist arrivals in destinations, associated to problems of crowding, localized inflation and pressure on housing, over time would in fact reduce the desirability of a tourism-led growth process. Understanding the social impacts of tourism on communities is therefore extremely important for governments to prevent a sentiment of hostility of local population against tourists and tourism development itself (Deery et al., 2012).

Thus, informed and collaborative destination management is essential for tourism destinations to be viable in the long term. In this regard, more sustainable tourism pathways in terms of environmental, social and economic effects require two main conditions. The first is the involvement of all stakeholders in tourism decision-processes and, in particular, the participation of resident communities (Choi & Sirakaya, 2005; Diedrich & García-Buades, 2009; McGehee & Andereck, 2004; Nunkoo et al., 2013). The second is the availability of basic information on the determinants of tourism expansion and on all the effects the sector may exert on destinations. Measuring and monitoring these aspects contributes to a more effective planning for a long-run sustainable tourism. For too long local policymakers have relied on a limited range of statistics, such as visitor-arrival numbers, employment surveys and visitor-satisfaction ratings, to monitor tourism in their destination. These statistics do not complete the description on the role played by tourism in a given destination. Collecting data and information on a broader range of issues, from the local economy to the community and the surrounding environment, will help policymakers and tourism strategists to have a more accurate description of both the potential and the real impact of tourism.

Besides being a catalyst of economic development, tourism can also work as stabilizer and binder for society. The existence of tourism is vital for the stability and development of society, especially in less developed regions/areas. Tourism is a comprehensive sector with extensive upstream and downstream linkages, including agriculture, manufacturing, transportation, retail and other economic activities (Rosalina et al., 2015). Tourist flows can increase the demand for goods and services (Rosalina et al., 2015) and promote the synergistic development of relevant industries, which translates into increased income for residents. It is also a labour-intensive sector, which provides numerous job opportunities for unskilled workers, as well as improving the quality of life of residents. An increase in tourism activities, indeed, promotes the growth of the economy, creates more job opportunities for residents and may also contribute to decrease crime rates (Ahad et al., 2021; Gould et al., 2002).

For all these driving factors, it is crucial to understand which processes and interventions contribute most to the attractiveness of a destination. This deeper investigation would require a shift from the preservation to the valorization of destinations, thus arising new challenges for a requalification of the supply side of the tourism industry. Heritage, as an example, is currently considered one of the most attractive factors and, therefore, under increasing pressure due to visitor flows. It has been estimated that "cultural tourism accounts for 40% of all European tourism; 4 out of 10 tourists choose their destination based on its cultural offering" (European Commission. Cultural Tourism), and "it is safe to assume that majority of tourist attractions and destinations in the world today are based on elements of cultural heritage" (Timothy, 2011, p. 3).

Some recent research (i.e. Rueda Márquez de la Plata et al., 2022) investigate how to improve the performance of both the tourism sector and the cultural heritage sector in a balanced and sustainable way, by considering the principles of conservation and preservation. Non-invasive tools such as ground penetrating radar (GPR), unmanned aerial vehicle (UAV) and even virtual reality (VR) and augmented reality (AR) technologies have been used to develop new methodologies, allowing us to generate new experiences and heritage tourist attractions, which not only do not generate negative impacts on the monuments themselves, but also promote their preservation without diminishing the cultural and tourist offers of the city.

These aspects lead to rethinking the traditional tourism paradigm, already severely stressed by the pandemic emergency; actually, there are other emerging values to focus on: such as, for example, the sustainability, the local stakeholders' engagement, urban regeneration, civic wealth.

In this scenario, the academic and managerial debate has widely discussed the importance of improving the innovation capacity of cultural heritage tourism so that it can be both more sustainable (Tu, 2020) and accessible to the whole community.

The availability of more detailed data is fundamental to help local destination coordinators to make informed decisions to improve tourism and destinations' performances. In this regard, the enriched ETIS toolkit would enable policymakers to monitor destination attractiveness and competitiveness and plan actions that, by exploiting the strengths and counteracting the critical issues of the area, will enhance the visitor experience, customers satisfaction while preserving/improving resident communities' quality of life. Inclusion, accessibility and innovation are the cornerstones upon which the sustainable development of the city of the future is based, in which cultural heritage must be the priority focus to stimulate processes of knowledge, valorization and participation (Kosmas et al., 2020).

Appendix

See Table A.6.1.

Table A.6.1 ETC/ULS proposal for TOUERM indicators

DPSIR[7] scheme components	TOUERM indicators	Indicator data sets
Drivers	Tourism flows	Tourism arrivals Overnights spent at tourism accommodation establishments Seasonality of tourism
	Tourism-related modes of transport	Tourism-related modes of transport: number of trips Tourism-related modes of transport (I): Airplane Tourism-related modes of transport (II): Cruises
	Most attractive places	Most attractive places
Pressures	Tourism density and intensity	Tourism density Tourism intensity Occupancy rate in tourist accommodation establishments
	Tourism pressure on protected areas	Tourism pressure on protected areas
	Water abstraction by tourism	Water abstraction by tourism
State	Bathing water quality	Bathing water quality reporting

(continued)

[7] DPSIR indicates driver–pressure–state–impact–response.

Table A.6.1 (continued)

DPSIR scheme components	TOUERM indicators	Indicator data sets
Impacts	Spatial impacts of tourism facilities	Spatial impact of tourism facilities (I): Golf courses Spatial impact of tourism facilities (II): Marinas Spatial impact of tourism facilities (III): Ski resorts
Responses	Percentage of destination that is designated for protection Tourism certification tools	Percentage of destination that is designated for protection Tourism enterprises using environmental certification/labelling (EMAS, EU Ecolabel, European charter for sustainable tourism) Blue Flags for beaches and marinas

Source Giulietti et al. (2018)

References

Ahad, M., Anwer, Z., & Ahmad, W. (2021). Does crime-tourism nexus hold for Pakistan? *International Journal of Emerging Markets, 17*(8), 1815–1839.

Balaguer, J., & Cantavella-Jordá, M. (2002). Tourism as a long-run economic growth factor: The Spanish case. *Applied Economics, 34*(7), 877–884.

Bonilla, J. (2023). *Achieving sustainable development goals through tourism: Toolkit of indicators for projects (TIPs)*. Instituto de Investigaciones en Turismo, Facultad de Ciencias Económicas, Universidad Nacional de La Plata.

Choi, H. S. C., & Sirakaya, E. (2005). Measuring residents' attitude toward sustainable tourism: Development of sustainable tourism attitude scale. *Journal of Travel Research, 43*(4), 380–394. https://doi.org/10.1177/004 728750527

Deery, M., Jago, L., & Fredline, L. (2012). Rethinking social impacts of tourism research: A new research agenda. *Tourism Management, 33*(1), 64–73.

De Siano, R., & Canale, R. R. (2024). The role of tourism in European regions' economic recovery: A spatial perspective. *Tourism Economics*, 13548166241248679.

Diedrich, A., & García-Buades, E. (2009). Local perceptions of tourism as indicators of destination decline. *Tourism Management, 30*(4), 512–521.

Dritsakis, N. (2012). Tourism development and economic growth in seven Mediterranean countries: A panel data approach. *Tourism Economics, 18*(4), 801–816.
Durbarry, R. (2004). Tourism and economic growth: The case of Mauritius. *Tourism Economics, 10*(4), 389–401.
European Commission. 2010. Europe 2020. A strategy for smart, sustainable and inclusive growth. COM(2010), *Brussels*, 2010.
European Commission. *Cultural Tourism.* https://ec.europa.eu/growth/sectors/tourism/offer/cultural_en
European Commission. (2016). *The European Tourism Indicator System ETIS toolkit for sustainable destination management.* Luxembourg.
European Commission. (2019). *European Tourism Indicators System for sustainable destination management.* http://ec.europa.eu/growth/sectors/tourism/offer/sustainable/indicators_en
Fahimi, A., Saint Akadiri, S., Seraj, M., & Akadiri, A. C. (2018). Testing the role of tourism and human capital development in economic growth. A panel causality study of micro states. *Tourism Management Perspectives, 28,* 62–70.
Giulietti, S., Romagosa, F., Esteve, F., & Schröder, C. (2018). Tourism and the environment. Towards a Reporting Mechanism in Europe.
Gould, E. D., Weinberg, B. A., & Mustard, D. B. (2002). Crime rates and local labor market opportunities in the United States: 1979–1997. *Review of Economics and Statistics, 84*(1), 45–61.
Hall, C. M., & Williams, A. M. (2008). *Tourism and innovation.* Routledge.
Kosmas, P., Galanakis, G., Constantinou, V., et al. (2020). Enhancing accessibility in cultural heritage environments: Considerations for social computing. *Universal Access in the Information Society, 19,* 471–482.
McGehee, N. G., & Andereck, K. L. (2004). Factors predicting rural residents' support of tourism. *Journal of Travel Research, 43*(2), 131–140.
Miralles, P. (2010). Technological innovation, a challenge for the hotel sector. *The European Journal for the Informatics Professional, 11*(2), 33–38.
Nunkoo, R., Smith, S. L., & Ramkissoon, H. (2013). Residents' attitudes to tourism: A longitudinal study of 140 articles from 1984 to 2010. *Journal of Sustainable Tourism, 21*(1), 5–25.
Razafindravelo, H. V. (2017). *Innovation strategies in tourism industry* (Master's thesis). University of Stavanger, Norway.
Rosalina, P., Garces, L. P. D. M., & Purisima, A. N. C. (2015). Tourism and crime: Evidence from the Philippines. *Southeast Asian Studies, 4*(3), 565–580.
Rueda Márquez de la Plata, A., Cruz Franco, P. A., Ramos Sánchez, J. A. (2022). Architectural survey, diagnostic, and constructive analysis strategies for monumental preservation of cultural heritage and sustainable management of tourism. *Buildings, 12*(8), 1156.

Timothy, D. J. (2011). *Cultural heritage and tourism: An introduction*. Channel View Publications.

Tu, H. M. (2020). Sustainable heritage management: Exploring dimensions of pull and push factors. *Sustainability, 12*(19), 8219.

Index

A
Agenda 2030, 5, 36
Air transport density, 84
Artificial intelligence, 124

B
Balaguer, J., 80
Biodiversity, 19, 20
Butler, R.W. (1980), 61, 80

C
Cantavella-Jorda, M., 80
Carbon emission intensity, 47
carrying capacity, 2, 63, 81
certification system, 118
City tourism congestion index, 85
climate change, 3, 17, 38, 41–45, 48, 101, 120
congestion, 4, 15, 36, 38, 57, 58, 60, 64, 67, 80, 82, 83, 85, 88, 91, 113, 117
controversial effect of tourism, 79
Cook, Thomas, 11

COP, 48
CSR, 120
cultural heritage, 27, 57, 94, 119, 122, 126, 128

D
decarbonization, 41, 50
digital transformation, 123

E
ECO-label, 120
economic growth, 4, 13, 15, 16, 18, 22, 34, 80, 81, 117, 122, 125
economic resilience, 4, 95, 97, 98
Economic sustainability, 17
ecotourism, 101, 121
EDEN, 120
EMAS, 120, 129
Emissions, 40
Environmental Action Programme, 121
Environmental Kuznets Curve, 22, 23, 122

environmental protection, 3, 41–43, 51, 60, 64
environmental sustainability, 3, 17, 34, 36, 37, 42, 44, 46, 49, 50, 62, 116, 117, 120, 122
ETIS, 4, 118, 119, 121, 128
European Charter for Sustainable Tourism, 58
European Environment Agency, 121
Eurostat, 47, 87–91, 99, 100, 111–114, 117
evolution of tourism, 7

G
Glasgow Declaration, 36, 49, 50
Glasgow Declaration on Climate Action in Tourism, 36
Gössling, S., 24, 34, 35, 41, 48, 51
Grand Tour, 10
Gross utilization index, 84

H
heritage tourism, 9, 128
history of tourism, 9, 10
Hospitality, 125

I
ICTs, 123
Industrial Revolution, 10, 11, 19
innovation, 3, 4, 44–48, 51, 115, 122, 123, 128

J
Johnston, R.J., 15, 69, 101, 102

L
life cycle approach, 61

N
Natural resources, 20

O
One Planet Sustainable Tourism Programme, 35, 49
overtourism, 4, 14, 35, 79–82, 84–91, 113

P
poverty, 3, 18, 41–43, 117
poverty reduction, 41–43

Q
quality of life, 14, 18, 20, 27, 43, 56, 57, 59, 62, 64, 67–69, 73, 75, 80, 82, 83, 103, 117, 127, 128

R
Resilience, 92, 93, 99, 100
Resilience index, 99

S
seasonality, 26, 62, 81–83, 116
sensitivity index, 98
Share of tourism in the economy, 85
short-stay accommodation, 85, 89, 90, 111, 112
smart destination, 123
Social sustainability, 17
supply chain, 44, 118, 120
sustainable development, 1, 4, 5, 16–18, 37, 38, 45, 56, 118, 120, 121, 128
sustainable tourism, 3, 4, 27, 28, 47, 58, 59, 64, 70, 101, 110, 118, 120, 125, 126, 129

T
Tourism, 7
Tourism activities, 35
Tourism Area Life Cycle, 13, 61
Tourism density, 82, 87, 129
tourism development, 2, 13, 27, 28, 58, 110, 115, 120, 126
tourism-growth nexus, 1, 13
tourism infrastructures, 11, 28
tourism innovation, 45, 46
Tourism intensity, 82, 88, 129
tourism policy, 114, 115
tourism resources, 4, 57, 70, 73, 76, 92, 101, 102, 117
tourism specialization, 15
tourism sustainability, v, 2, 55, 56, 58, 63, 65, 74, 102, 113, 116–119
Tourism territorial pressure, 83, 89
Tourism Working Group, 49

tourist accommodation, 111, 129
tourist arrivals, 12, 15, 40, 88, 110, 111, 126
transports, 38, 39
Tyrrell, T.J., 15, 69, 101, 102

U
UN, 7, 27, 36, 46, 48, 50, 121
UNEP, 36
UNFCCC, 48
UNWTO, 39

W
waste, 17, 19, 36, 57, 60, 81, 82, 86, 117, 119, 120, 124–126
World Tourism Organization, 2, 14, 27, 121

9783031854842